Recent Advances in Welding Simulation

Papers presented at a one-day seminar *Recent Advances in Welding Simulation* on
26 November 1999 held at IMechE Headquarters, London, UK

IMechE
Seminar Publication

I MECH E

Recent Advances in Welding Simulation

Organised by
The Materials and Mechanics of Solids Group
of the Institution of Mechanical Engineers (IMechE)

IMechE Seminar Publication 2000–13

Professional Engineering Publishing

Published by Professional Engineering Publishing Limited for The Institution of Mechanical Engineers, Bury St Edmunds and London, UK.

First Published 2000

ISSN 1357–9193
ISBN 1 86058 310 5

A CIP catalogue record for this book is available from the British Library.

Contents

S708/001/2000	**Welding residual stresses and distortions simulated by the use of simplified methods** L-E Lindgren and B L Josefson	
S708/002/2000	**Modelling of residual stresses at girth welds in pipes** E Olden and R Leggatt	21
S708/003/2000	**Nickel-base superalloy welding – models for residual stress, distortion, and weldability** S M Roberts, O Hunziker, D Dye, H J Stone, and R C Reed	33
S708/004/2000	**New capabilities of the software product SYSWELD for the prediction of welding residual stresses and distortions** G Mangialenti and T Carter	45
S708/005/2000	**Measurement and prediction of residual stresses in a thick section steel welds** D J Smith, D George, P J Bouchard, and C Watson	55
S708/006/2000	**Friction stir welding of aluminium extrusions** L D Oosterkamp, O J Emmerhoff, and O T Midling	79
S708/007/2000	**Integrity of welded joints in pressure vessels** R K Penny and W Kohlhöfer	89
S708/008/2000	**A study of finite element meshing methods to be used in welding simulation for ship structures** A C Ramsay	99
Authors' Index		119

Related Titles of Interest

Title	Editor/Author	ISBN
IMechE Engineers' Data Book – Second Edition	C Matthews	1 86058 248 6
Handbook of Mechanical Works Inspection – A Guide To Effective Practice	C Matthews	1 86058 047 5
Integrity of High-Temperature Welds	IMechE/IOM	1 86058 149 8
Stress Analysis of Cracks Handbook	H Tada, P C Paris, and G R Irwin	1 86058 304 0
Remanent Life Prediction	IMechE Seminar	1 86058 154 4
Fracture Mechanics of Welds (EGF 2)	J G Blauel and K H Schwalbe	0 85298 644 0
Mis-Matching of Welds (ESIS 17)	K H Schwalbe and M Kocak	0 85298 898 2

For the full range of titles published by Professional Engineering Publishing contact:

Sales Department
Professional Engineering Publishing Limited
Northgate Avenue
Bury St Edmunds
Suffolk
IP32 6BW
UK

Tel: +44 (0)1284 724384
Fax: +44 (0)1284 718692

S708/001/2000

Welding residual stresses and distortions simulated by the use of simplified methods

L-E LINDGREN
Luleå University of Technology, Sweden
B L JOSEFSON
Chalmers University of Technology, Göteborg, Sweden

Abstract

Finite element modelling of the generation of welding residual stresses and distortion is reviewed. Focus will be put on continuum mechanics based constitutive modelling, that is models suitable for simulation of residual stress generation on a macro-scale, techniques for a good finite element modelling generating residual stresses. Two applications will be studied in more detail: residual stresses in built-up spot welded car body components and multi-pass butt welding of thick-walled pipes and plates.

1 Introduction

Welding residual deformations and stresses will be generated in a structure as a consequence of local plastic deformations introduced by local temperature history consisting of a rapid heating and subsequent cooling phase. The residual stress field is important for the function of the structure as it may influence the mechanical behaviour of a structure in several ways. Simulation of welding is increasingly used as a tool for design of welding processes and assessment of the performance of welded components. The first aspect is usually concerned with welding deformations whereas the latter is normally concerned with residual stresses. The transient deformations which change the gap in the groove may affect the welding process itself leading to either a burn-through or a too small penetration. The residual deformations can give problems with shape misalignments in the assembly process and increase the risk for buckling in welded build-up structures. Okumoto and Matsuzaki [1] study the measured accumulated error in geometry in the assembly of ship hull components. They quantify statistically determined errors resulting from different steps in the assembly process. To improve the accuracy in the production process the residual deformations caused by cutting and welding were calculated using a detailed non-linear FE-analysis. The effect of welding sequence on panel distortion an a built-up structure i studied by Tsai et al. [2].

To quantify the effects of welding residual stresses in the design stage, one has to numerically simulate the redistribution, and possible relaxation, of residual stresses in a structure subject loaded by design stresses. This underlines the need for robust, reliable and computer efficient methods of modelling residual stresses. Tensile residual stresses are detrimental for the initiation and growth of fatigue cracks and welds are often particularly sensitive to fatigue due to the presence of stress concentrations introduced by the weld imperfections. Moreover, in the weld region, the welding residual stress will add to the applied stress range, which will increase the active stress range opening the crack. Hence, crack growth rate will be considerably higher in welds.

The influence of residual stresses on the fatigue life is studied numerically in [3-6] modelling the residual stress field after welding with different levels of accuracy. Static growth of cracks in welding residual stress fields is studied in [7-10] based on the interaction of the numerically calculated through-the-thickness variations of the residual stress component opening the crack and an external design load. Compressive residual stresses together with shape misalignments and global residual deformations affect the buckling strength of a manufactured structure. Michaleris and deBiccari [11], and Michaleris and Sun [12] compare numerical and experimental predictions of the buckling load for panels having welded stiffeners with the aim to find the influence of heat input and panel geometry on the buckling distortion caused by membrane welding residual stresses. Note also that membrane residual stresses may influence the dynamic response For example, residual membrane stresses in thin-walled structures will alter the resonance frequencies and eigenmodes, see [13].

One may distinguish between three different approaches to model mechanical effects of welding [14-15]. Their applicability depends on the resources available (time, expertise and tools) and the scope of the investigation. The latter means what questions are to be answered and what accuracy is needed. In general, the nearer the weld and the higher the temperature are where the thermal and mechanical fields are wanted, the more accurate models are required. The three different approaches are:

- Empirical methods based on experiments or analytical solutions
- Numerical methods employing a simplified temperature history and field
- Numerical simulations employing detailed nonlinear FE-analyses

2 Empirical methods based on experiments or analytical solutions

Analytical solutions for welding residual distortions and stresses are collected in [16] and more recently by Leggatt [17]. Even if residual stresses can be fairly well modelled one should validate the calculations. This motivates the measurement of residual stresses. One usually distinguishes between destructive and non-destructive methods of measuring residual stress, see [18].

3 Numerical methods employing a simplified temperature history and field

It is found that the welding residual distortion, and also residual stresses, is determined by the local plastic "shrinkage" strains generated during the cooling after the welding. Hence, although the weldment undergoes many complex physical changes, involving the interaction with between microstructure, thermal and mechanical fields (see below), during the welding process, the plastic strains accumulated during the final stages of cooling largely determine the residual distortion. This fact is employed by several simplified methods, and is sometimes referred to as a "inherent shrinkage model". This also means that the temperature and corresponding strain history present during the early stages of welding can be neglected. However, it is important to remove plastic strains accumulated in material when heated to temperatures above the melting temperature followed by subsequent cooling, that is for material in the HAZ close to the weld metal, see [19], otherwise the welding residual distortion may be much smaller than the corresponding measured distortion. Tsai et al. [2] use this technique to calculate the distortion in stiffened panels using FEA. The plastic strains, or rather the residual stress and strain fields, generated in one weld is then calculated by cooling the weld area from a prescribed high temperature. The use of a simplified temperature field and history is also discussed in Josefson [15] where the welding process is modelled as a linear (in time) heating to the temperature subsequent linear cooling of material in the weld. Two applications are studied: Multi-pass butt welding of a pipe [20], where the temperature shape for material outside the weld is calculated using analytical methods, and spot welding [13], where material outside the spot is kept at ambient temperature. In both cases qualitatively good residual stress shapes are obtained. In Josefson et al. [21], the residual deformations for a ring-stiffened pipe is determined using a simplified model where the temperature history during welding is considerably simplified, see [20], which enables a quick answer for the residual deformation after the deposition of several weld passes.

4 Numerical simulations employing detailed nonlinear FE-analyses

There exist several commercial FE-codes that may be employed for detailed nonlinear simulations of the development of the temperature and stress fields present during manufacturing processes like welding. Wilkening and Snow [22] show, in detail, how to use one such general FE-code, ADINA, for a welding mechanics analysis. These codes may not allow the user to account for all nonlinearities and couplings between different fields considered in research codes. However, they do have a number of options for type of finite element formulation and procedures to account for the thermo-mechanical couplings. Some codes do also offer the possibilities to enhance their capacity by user defined routines. An outline of the different ways thermo-mechanical couplings can be handled are given below together with choices of finite element formulations. Furthermore, the quality of the FE-simulations depends strongly on the constitutive modelling of the material behaviour. General thermo-elastoplastic constitutive models are discussed together with additional aspects that are needed in welding simulations. Two applications where detailed FEA-analyses are used are also given.

5 Accounting for couplings between different fields

The welding process is complex. There is currently no complete model accounting for the arc physics, the weld pool phenomena, and finally the deformation and heat conduction in the solid metal. Modelling spot welding requires the inclusion of the electrical field [6, 23]. Preheating or post weld heat treatment by induction heating also involves the magnetic field. There are also models that includes the hydrogen diffusion. We will in the following focus on the thermo-mechanical fields, Figure 1. The coupled thermo-mechanical problem is illustrated in Figure 1. Simulations that are concerned with the mechanical effects of welding have to compute the thermal and mechanical fields. The material behaviour, due to the changing microstructure, may depend on the temperature and deformation histories. This requires the consideration of the microstructure evolution when modelling the material behaviour. The possible couplings are shown in Figure 1 and explained in Table 1. Some of the couplings are usually ignored. Most analyses ignore the effect of mechanical generated heat which is small compared to the heat input in the weld, couplings 4b-d. It is also often possible to neglect the influence on the changing geometry on the thermal behaviour, coupling 4a. An exception is the case when the contact with fixtures near the weld are changed due to the deformation or in the case of spot welding when the changing contact area is very important for the thermal-electrical fields. The effect of stresses on phase changes are also ignored, coupling 6, and in many cases the phase changes are ignored completely.

Then a weakly coupled simulation can be performed, Figure 2a. The thermal analysis is performed first. The thermal properties includes some effects of known phase changes like latent heats etc. Thereafter the mechanical analysis is performed and likewise some effects of known phase changes are incorporated. However, a combined simulated using the staggered approach, Figure 2b without the micro structure model, is more practical than two separate simulations. Then it is straightforward to include coupling 4 and 6.

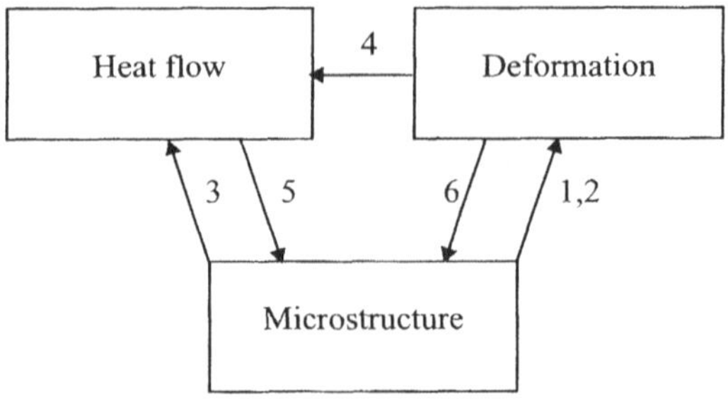

Figure 1. Phenomena occurring during welding.

Table 1. Couplings in Figure1.

Coupling #	Explanation
1a	Thermal expansion depends on microstructure of material.
1b	Volume changes due to phase changes.
2a	Plastic material behaviour depend on microstructure of material.
2b	Elastic material behaviour depend on microstructure of material.
3a	Heat conductivity and heat capacity depend on microstructure of material.
3b	Latent heats due to phase changes.
4a	Deformation changes thermal boundary conditions.
4b	Heat due to plastic dissipation (plastic strain rate).
4c	Heat due to thermal strain rate.
4d	Heat due to elastic strain rate.
5	Microstructure evolution depends on temperature.
6	Microstructure evolution depends on deformation.

However, there can be a problem of finding models and properties for the latter influence. It is necessary to account for the phase changes when simulating welding of ferritic steels. The phase changes mean that the material properties do depend on the temperature and the temperature history. This can be handled by the approach in Figure 2a for a one pass weld [24]. However, a model for the microstructure evolution, Figure 2b, and the pertaining material properties are needed for multipass welds of ferritic steels

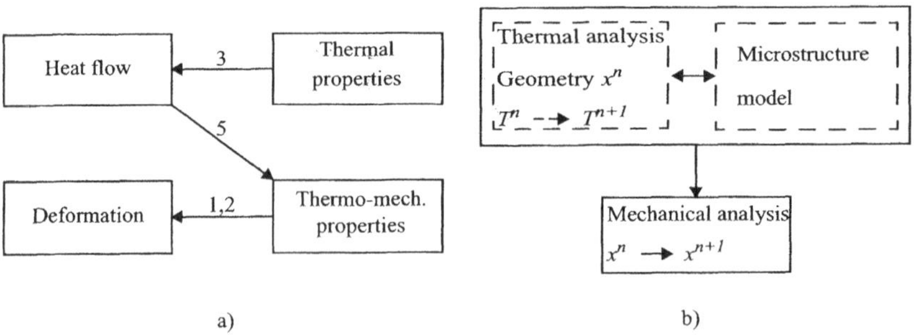

Figure 2. A weakly coupled analysis is shown in a) and a time step in a staggered approach, which can account for more couplings, is shown in b).

[25, 26]. It is possible to use a full coupling where the thermal and mechanical fields are solved simultaneously. This gives an unsymmetric matrix in the solution process. This approach is not to be recommended. If the coupling is very strong then a staggered approach with smaller time steps and iterations between the thermal and mechanical solution in a time step may be accurate enough.

5.1 Finite element analysis for the thermal field.

There are not so many options available in finite element analysis of heat conduction. This is probably due to the fact that these analyses are quite straightforward even if the thermal properties are temperature dependent. However, there are some different approaches developed in order to handle the numerical problems that may exists during phase changes due to latent heat effects. The solution procedure can use a so-called effective heat capacity [27] or the fictitious heat source [28]. Still there will be problems in the case of a pure metal where the melting/solidifying occurs at one temperature. Then the user most regularize the latent heat by spreading it over a larger temperature change. The most accurate approach, which can even handle the pure metal case, is an enthalpy formulation. It is based on a reformulation of the heat conduction equation into enthalpy. Then enthalpy is the nodal unknown. It that can be translated into temperature upon output of the result. The thermal analysis is seldom solved using an explicit formulation, see below for mechanical analysis, due to the time step constraint is even more severe in this case as it is inverse proportional to the size of the smallest element in square. Usually a generalized midpoint method is used for the time stepping algorithm [27].

5.2 Finite element analysis for the mechanical field.

The solid, Lagrangian, formulation [29] is a straightforward choice when the aim is to obtain residual stresses although a flow, Eulerian, formulation has been effectively used for obtaining the steady solution for a moving arc in large plate [30, 31]. The next choice is whether to use an explicit or implicit formulation [32]. The explicit formulation can be more efficient in some cases and is easy to use for strongly nonlinear problems. However, as it is a formulation for dynamic problems, special precautions are necessary for obtaining static solutions. The implicit is recommended. The explicit method is conditionally stable and requires a very small time steps. Mahin [33] and coworkers have used explicit formulations for welding simulations. Other options has to do with the type of element [34]. Low order elements with underintegration of the volumetric strain are recommended. The latter is necessary in order to avoid locking due to large, incompressible plastic strains. The implicit method also requires the user to choose type of solver [35] in the solution of the global system of equations. The general algorithm for a nonlinear simulation is given in Table 2. Other options exist and but the basic approach is based on finding zero residual forces using a full Newton-Raphson method. Other methods, Quasi-Newton strategies etc., perform step 3 in Table 2 differently. The formulation of the tangent stiffness matrix $\mathbf{K}_{\text{eff}}$ determines the rate of convergence in the iterative procedure. To obtain second order rate of convergence in this procedure $\mathbf{K}_{\text{eff}}$ should be calculated using a tangent modulus matrix consistent with the linearization in the Newton-Raphson method. The computation of the residual load $\mathbf{R}$ is important for the accuracy of the finite element solution. This residual load requires the accurate computation of the stresses which is discussed later. Efficiency can be improved by using adaptive meshing [36-39] and parallel computing. It is not possible to deal with all these issues in this paper. We will focus on the local problem of stress computation in the following.

Table 2: Incremental solution with iterative corrections of nonlinear problems using the finite element method.

NONLINEAR FINITE ELEMENT ANALYSIS
Step 1. Initialize analysis Increment (time step) counter $n=1$ Initialize displacements, velocities etc.
Step 2. Next increment Iteration counter $i=1$ Initialize estimate of displacements, velocities for ${}^{n+1}t$. ${}^{n+1}_{i}U = {}^{n}U$
Step 3. Solve equations and update estimate for ${}^{n+1}t$ ${}_{i}K_{eff\,i+1}\Delta U = -{}_{i}R = -({}_{i}F_{int} - {}_{i}F_{ext}) \Rightarrow {}^{n+1}_{i+1}U = {}^{n+1}_{i}U + {}_{i+1}\Delta U$ where ${}_{i}F_{int}$ = assembly of $f_{int} = \int_{V_e} B_e^T \sigma dV$
Step 4. Check convergence If no convergence then $i=i+1$ and go to step 3 If convergence then ${}^{n+1}U = {}^{n+1}_{i+1}U$ and $n=n+1$. If more increments and go to step 2.

5.3 Modelling of mechanical behaviour of material

It is important to have correct material models and corresponding accurate numerical methods for integrating these stress-strain relations. Errors in this part of a finite element model can not be compensated by increased accuracy in time stepping, finite element discretisation etc. The material is experiencing a large variation in temperatures and strain rates during a thermal cycle. Furthermore, as mentioned earlier, the effect of the phase changes makes the modelling even more complex. It is necessary to simplify the high temperature behaviour of the material both because of lacking data and numerical problems associated with the material behaviour at high temperatures.

Some preliminaries about large strains are followed by a general outline of plasticity and accurate numerical solution of stress-strain relation. The section is concluded by some remarks specific for the modelling of welding.

Large strain formulation. Simulation of welding requires a large deformation analysis in most cases. It is no large additional cost to account for both large strains and deformations. An additive decomposition of strain increments is assumed or derived based on a multiplicative decomposition of the deformation gradient [40]. One such approach of extending small strain plasticity to finite strains is demonstrated in [41]. There still remains, however, some open questions in this respect [42-44]. For cases where the elastic strain is much smaller than the plastic strain one may simplify the decomposition of the deformation gradient to an additive split of the velocity strain, that is a hypo-elastic constitutive law, where a suitable stress rate is proportional, via Hookes law, to the velocity strain tensor (or formally to the total velocity strain minus the plastic velocity strain) [43]. It is essential to employ a so called objective stress rate in the hypo-elastic constitutive law above [44]. The hypo-elastic relations are the basis for applying the constitutive model and the stress computation at a given reference configuration where initial stresses, strains and incremental quantities all are related to the same configuration. The algorithm, for computing stresses will then be form-identical to the small strain case [45]. The following presentation will therefore only describe stress and strain relations without any reference to large or small strains.

Plasticity. The plastic strains cause the residual stresses. They are modelled in varying ways even if the different models correspond to the same physical processes. This is the creation, annihilation etc. of dislocations. However, it is convenient to use different models depending on the time scale, the temperature and stress range in focus. The main division is rate (time) dependent and rate (time) independent plasticity. All processes require time but this can be ignored in some applications and then rate independent plasticity is used. A typical plasticity model consists of a yield condition, flow rule and a hardening rule. The yield condition, if there is a domain with pure elastic behaviour, defines the boundary between elastic and plastic behaviour. The stress state must stay on or inside this yield surface for rate independent plasticity. The flow rule gives the equations for the plastic strains in terms of stresses. It is usually in rate-form. The hardening rule defines the hardening and possibly also softening behaviour of the material. The equations for von Mises plasticity are given in the next section. The large temperature range makes it necessary to account for temperature dependent properties. The von Mises plasticity model is used with linear or non-linear hardening models. The hardening can be a combination of isotropic and kinematic hardening. The latter will influence the residual stress state near a weld even for a one pass weld due to the thermal cycle [46]. The algorithm for stress computation is outlined in the next section. Material subject to cyclic loading may either stabilize after several stress-strain loops, or a progressive increase in strain for each cycle is observed, so called ratchetting [47]. The implementation of material models for ratchetting into non-linear finite element codes suitable for welding mechanics analysis is described in[48].

Accurate integration of incremental strain-stress relations. The finite element simulation is performed as a time-stepping procedure. The increment in displacements corresponding to equilibrium at the end of a time step are computed. The stress at the end of the time step is needed in order to check equilibrium. Thus the increment in displacements give increment in strains. It is necessary to compute the corresponding stress increment for the actual time step. It is important to evaluate the stress increment accurately as the solution will otherwise not correspond to equilibrium.

The effective stress function is a generalisation of the radial return method. It was named by Kojic and Bathe [49]. It is an accurate and effective numerical procedure for computing stresses from given strain history. The main problem is to find the amount of elastic and inelastic strain increments during a time step. It is formulated as a nonlinear scalar equation for the effective stress. It is a quite general method applicable to combinations of rate independent and rate dependent plasticity. Mixed, nonlinear kinematic and isotropic hardening can be accounted for. It is straight forward to include phenomena like transformation plasticity [50] into the algorithm. It can also be extended to treat compressible plasticity, that is yield functions with a mean stress dependency [51].

The algorithm is summarised in Table 3. All relations are set up in incremental form. We use the notation state A for the given initial state at the beginning of a time step and state B for a trial state. The trial state is computed by assuming that the increment from state A is elastic. We denote the wanted final state at the end of the time step with C. The basic idea is simple. The stress state, and therefore also the effective stress, depends on the amounts of elastic and inelastic strains. The effective stress must be on the yield surface at state C for rate-independent plasticity. The solution is obtained by "guessing" the final effective stress and computing the corresponding inelastic strains and yield limit. This in turn make it possible to compute the yield limit. The process is repeated until the effective stress and yield limit are equal for state C. A more mathematical formulation is given below. The algorithm is modified if no yield surface exists. This is the case for pure visco-plasticity. Then the flow rule (flow strength equation) plays the role of the "yield surface" in the effective stress function logic [51].

The hydrostatic part is a straightforward updating as we assume incompressible plastic, behaviour and the thermal strain is given. Hooke's law then gives

$$_m\sigma_C = K(T_C)\left(\frac{_m\sigma_A}{K(T_A)} + 3(\Delta_m\varepsilon - \Delta\varepsilon_t)\right) \quad (1)$$

where

$_m\sigma_A$ and $_m\sigma_C$ are the hydrostatic stresses at the beginning and the end of the time step,

$K(T_C)$ is the bulk modulus for the temperature at the end of the time step,

$\Delta_m\varepsilon$ is the volumetric total strain (known), and

$\Delta\varepsilon_t$ is the thermal strain (known).

The relation between deviatoric stresses and strains is

$$s_C = 2G(T_C)\left(\frac{_ms_A}{GT_A} + \Delta e - \Delta e_p - \Delta e_c - \Delta e_{tp}\right) = 2G_C\left(e^e{}_A - \Delta e_p - \Delta e_c - \Delta e_{tp}\right) \quad (2)$$

where

s_A and s_C are the deviatoric stresses at the beginning and the end of the time step,

$G(T_C)$ is the shear modulus for the temperature at the end of the time step,

$e^e{}_A$ is the elastic deviatoric strains at the beginning of the time step,

Δe_p is the increment in plastic strains,

Δe_C is the increment in creep strains, and

Δe_{tp} is the increment in transformation plasticity strains.

Many creep laws state that the creep rate is proportional to the deviatoric stresses as

$$\Delta e_c = \Delta t\gamma_C s_C \quad (3)$$

The parameter γ depends on e.g. the material parameters and the stress state. The transformation plasticity is written as [50], assuming diffusion driven isotropic transformations,

$$\Delta e_{tp} = \Delta t\phi_C s_C \quad (4)$$

The parameter ϕ depends on material properties and phase transformation rate. Use of von Mises yield criterion, the associated flow rule and assuming that the direction of flow is determined by the final stress state give

$$\Delta e_p = \Delta\lambda\frac{3}{2\bar{\sigma}_C}s_C \quad (5)$$

where

$\Delta\lambda$ is the wanted parameter determining the magnitude of the plastic flow.

The effective plastic strain increment will be

$$\Delta\bar{\varepsilon}_p = \Delta\lambda \quad (6)$$

We rewrite the sum of (3) and (4) as

$$\Delta e_c + \Delta e_{tp} = \Delta t\gamma_C s_C + \Delta t\phi_C s_C = \Gamma_C \frac{s_C}{2G_C} \text{ with } \Gamma_C = 2G_C(\Delta t\gamma_C + \Delta t\phi_C).$$

Equations (3-5) and the equation above are insert into Equation (2) giving

$$s_C = 2G_C\left(e^e{}_A - \Delta\lambda\frac{3G_C}{\bar{\sigma}_C}s_C - \Gamma_C\frac{s_C}{2G_C}\right) \Rightarrow s_C\left(1 + \Delta\lambda\frac{3G_C}{\bar{\sigma}_C} + \Gamma_C\right) = 2G_C e^e{}_A$$

The effective stress is finally obtained as

$$\bar{\sigma}_C = 2G_C\frac{\sqrt{\frac{3}{2}}\left\|e^e{}_A\right\|}{\left(1 + \Delta\lambda\frac{3G_C}{\bar{\sigma}_C} + \Gamma_C\right)} \Rightarrow \bar{\sigma}_C = \frac{2G_C\sqrt{\frac{3}{2}}\left\|e^e{}_A\right\| - 3G_C\Delta\lambda}{(1+\Gamma_C)} \tag{7}$$

A hardening rule gives the yield limit at state C. Assuming linear, isotropic hardening gives

$$\sigma_{yC} = \sigma_{yB}(T_C) + \Delta\varepsilon_p H' = \sigma_{yB}(T_C) + \Delta\lambda H' \tag{8}$$

The yield limit $\sigma_{yB}(T_C)$ is the yield limit that will exist at state C if the increment is elastic.

Setting the effective stress at state C, Equation (7), equal to the yield limit, Equation (8), gives,

$$\Delta\lambda = \frac{2G_C\sqrt{\frac{3}{2}}\left\|e^e{}_A\right\| - (1+\Gamma_C)\sigma_{yB}(T_C)}{((1+\Gamma_C)H'_C + 3G_C)}, \tag{9}$$

Thus, if the stress at state C is known, then one can compute the plasticity parameter in Equation (9) and the inelastic strains in Equations (3-5). Equations (8) and (9) will then give the yield limit at state C. The consistency conditions requires that this yield limit should be equal to the effective stress. Setting the effective stress in Equation (7) equal to the yield limit in Equation (8) is the so-called "Effective Stress Function". Note, that by formulating the problem in terms of the effective stress, a single equation for a scalar entity is obtained, which can be solved efficiently with standard method like numerical bisection methods. It is here solved in an iterative manner by trying different effective stress values until it is fulfilled. However, no iterations are required for the special case of von Mises plasticity and linear hardening. Then (9) is the exact solution for the plastic flow and the esf-logic corresponds to the radial return method. The solution procedure is outlined in Table 3 below.

Special consideration for simulation of welding. Many simplification in modelling of welding are concerned with the high temperature behaviour. See the review by Lindgren [52] for more details.

Nearly all analyses ignore the fluid flow, an exception is [53]. The actual weld behaviour is approximated with a hot, soft solid. A high conductivity is given the molten pool in order to imitate the stirror effect which gives convective heat transfer. The net heat input into this region must be found by comparison with experimental results. Most models neglects the viscous behaviour at high temperatures. This is possible because the time spent at high temperature is short and phase changes affect the dislocation structure during cooling and thereby reduced the influence of the processes at high temperature. It is important, as stated earlier, to remove the accumulated plastic strains when the material is melting [19, 54]. Furthermore, a so-called cut-off temperature is used. Temperatures higher than this value will be set to this value in the mechanical analysis.

Table 3: Algorithm for effective stress function.

STRESS COMPUTATION
Compute estimate af amount of plastic flow
1st time (trial stress computation)
$\bar{\sigma}_1 = \bar{\sigma}_B \Rightarrow \gamma_1, \phi_1, \Delta\lambda_1 \Rightarrow err_1 = \bar{\sigma}_1 - \sigma_{y1}$
If no plastic yielding, creep or transformation plasticity then this is the final stress.
2nd time
$\bar{\sigma}_2 = \bar{\sigma}_A \Rightarrow \gamma_2, \phi_2, \Delta\lambda_2 \Rightarrow err_2 = \bar{\sigma}_2 - \sigma_{y2}$
Later times
Use e.g. interpolation method for generating new value for effective stress
$\bar{\sigma}_3 = \bar{\sigma}_2 - err_2 \dfrac{\bar{\sigma}_2 - \bar{\sigma}_1}{(err_2 - err_1)}$
$\Rightarrow \gamma_3, \phi_3, \Delta\lambda_3 \Rightarrow err_3 = \bar{\sigma}_3 - \sigma_{y3}$
Check convergence
If convergence then
$\bar{\sigma}_C = \bar{\sigma}_3$ Update stress components etc.
else
Keep two pairs of $\bar{\sigma}$ and *err* for generating a new estimate

These simplifications have been found to work also well in cases where the residual stresses are of interest [55]. Thus the material properties above this temperature do not influence the computed results.

Modelling of multipass welding poses challenges not only to the material modelling [26] for materials undergoing phase changes but also the modelling of the addition of filler material requires a special consideration. It is not so practical to restart and redefine the model for every new weld pass. The user wants to define the complete model at the start of the analysis. There are two basic approaches. The first, called quiet elements, has all elements and nodes present in the analysis. However, elements corresponding to not laid welds are given material properties so they do not affect the rest of the model. They are "quiet". It is also important to assure that they do not accumulate plastic strains during this stage that will affect their later behaviour. They are given correct material properties at the time the weld is laid. The other approach, inactive elements, does not include these elements in the analysis. They are defined by the user at the start of the analysis but the nodes in the interior of the weld are not part of the unknown nodal displacements. The elements and nodes are added to the analysis when the weld is laid. Then the profile of the finite element matrix must be recomputed. The first approach is easier to apply in a code whereas the latter method requires a code that has this model change capacity. It is shown by Lindgren et.al. [56] that both approaches can give the same results. The inactive method has problems in the case of large deformation. The the code must have a special capability for moving them to the correct location with respect to the deformed geometry.

6 Applications

6.1 Mechanical modelling of resistance spot welding

In resistance spot welding heat is generated by contact resistance as copper electrodes are clamped to the sheets being welded. An accurate model would then include the interaction of electrical, thermal and mechanical fields as well as the microstructure evolution. Henrysson et al. [6] modelled spot welding of a single overlap specimen using the commercial FE-code SYSWELD. The material is modelled as elastic-plastic with linear kinematic hardening. Differences in yield stress for the different microstructural phases were included. The residual stress field for one spot weld was calculated assuming rotational symmetry. Figure 3 shows the calculated residual radial stress variation on the outer surface of the spot weld together with corresponding residual stresses measured by use of the X-Ray method. High stress gradients were observed at the edge of the nugget both in the FE-analysis and in the experiments. The FE-calculated residual stress field was transferred to a fully 3D FE-model representing an overlap specimen. Fatigue loading was then simulated by a cyclic loading of the specimen. The flow chart for the study is shown in Figure 4. The largest radial stress amplitude due to mechanical loading is normally found to appear at the perimeter of the nugget. This was found also in [6], however as the yield stress of the nugget is about twice as high as in the base metal due to phase changes during welding, it was found that for higher membrane loads (with magnitudes normally found in car body structures) the material yields in each load cycle in the base metal (with the largest experienced plastic strains) whereas the material is elastic at the nugget perimeter during the cyclic loading. The radial stress is used for fatigue life predictions since cracks often initiate in the sheets, in the circumferential direction, outside the nugget. A similar approach for the residual stress calculation is presented by Sun et al. [23] using the FE-code ABAQUS together with a package of user-supplied routine to model the interaction with the electrical field. For built-up structures it is not possible to model the generation of residual stresses in each spot in detail. Alfredsson and Josefson[13] used simplified method above to model the influence of residual stresses on subsequent dynamic loading in a box beam welded with

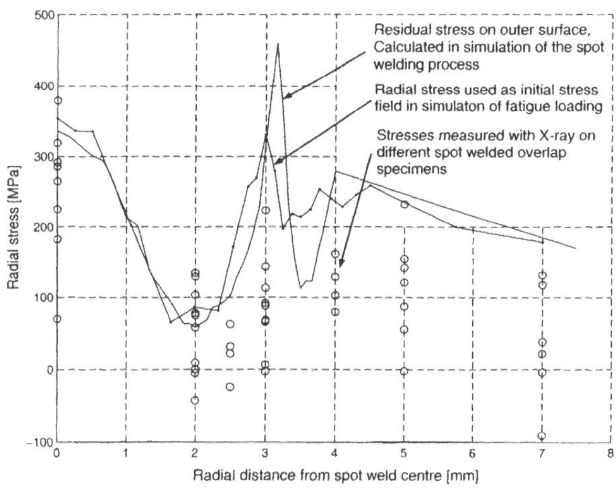

Figure 3. Radial stress on the outer surface of a spot weld. From [6].

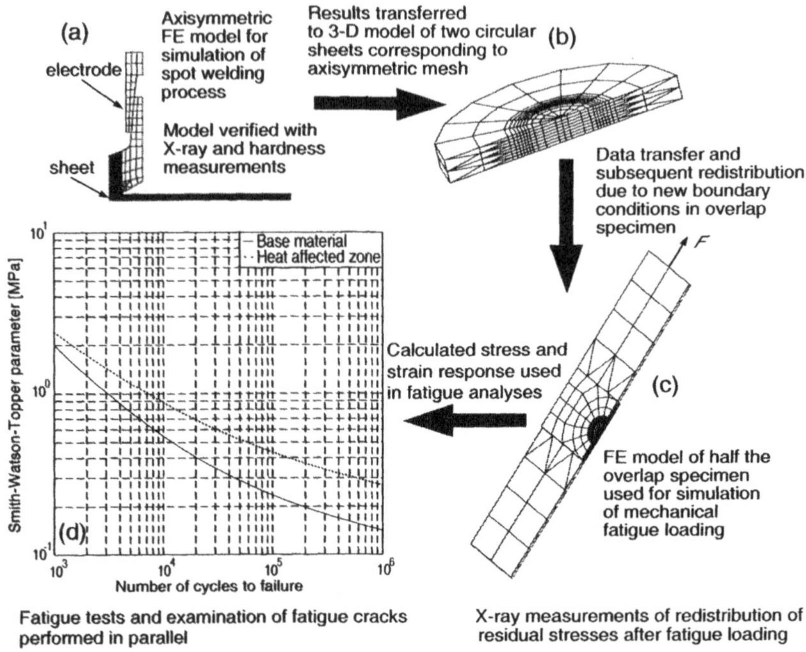

Figure 4. Flow chart showing the different stages of the investigation: (a) Simulation of spot welding process; (b) Data transfer to overlap specimen; (c) Simulation of fatigue loading; and (d) Fatigue analyses using calculated stresses and strains. From[6].

10 spots. Car body structures, or components of the car body, contain many spots. In mechanical analyses they are normally modelled using beam elements, thus considering their contribution to the stiffness but normally neglecting effects of welding residual stresses.

6.2 Multipass weld of thick plate made of ferritic steel

The 2 dm thick plate was welded with submerged arc welding. The joint was prepared in such a way that an artificial crack appeared in the interior of the groove. The welding is simulated using a 2D-model. Thus the heat flow in the welding direction is neglected and only a cross-section is analysed. Plane strain is assumed in the mechanical model. See the paper by Börjesson and Lindgren [26] for details. Part of the finite element model is shown in Figure 5 where the individual weld passes also have been denoted. The microstructure evolution was computed simultaneous with the temperatures. This information is used to compute the thermal and material properties during the simulation. Each phase (Martensite, Bainite, Pearlite, Ferrite and Austenite) is assigned material properties versus temperatures. These are combined using a simple linear mixture rule for computing the property of the material. Furthermore, latent heats and volume changes associated to the phase changes are also computed. An example of computed yield limit is given in Figure 6. The measured and computed residual transverse stresses at the top of the plate are shown in Figure 7.

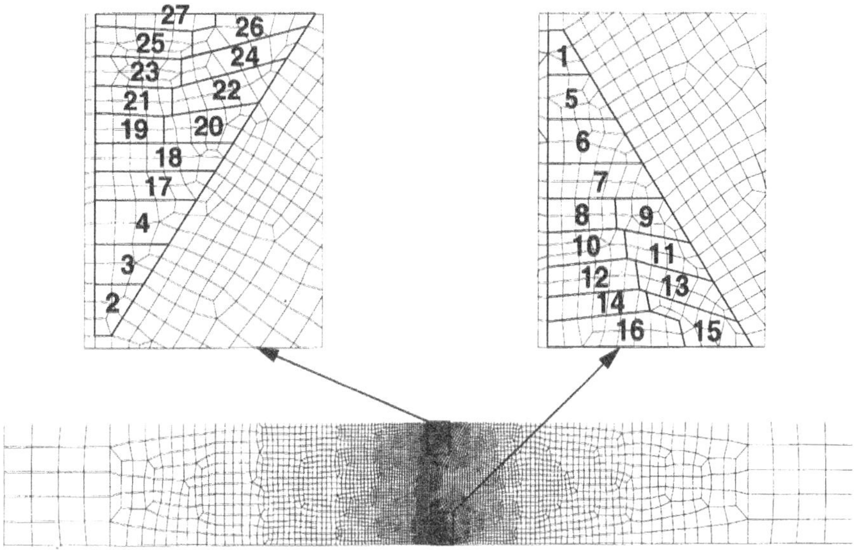

Figure 5. Finite element model o multipass weld. From [26].

Simulations can be used as an additional tool in evaluating welding procedures for setting up a Welding Procedure Specification. They can give the residual stress fields which may be important for the long time integrity of the weld. Hedblom et.al. [57] compared a conventional U-seam weld with a type of narrow gap weld, Optigap. The welds were circumferential butt weld. The diameter of the welded pipes was 168.3 mm and the wall thickness was 11 mm and the pipes were made of AISI 316L. The welding was done by mechanized TIG with AISI316LSi filler material. The U-seam consisted of ten weld passes placed in five layers. One pass in the first layer then two passes per layer until the top layer with three passes. The groove was 12 mm at the outer surface of the pipe before welding. For the Optigap procedure the weld consisted of eight passes placed in seven layers, one weld pass per layer until the top layer with two passes. The Optigap groove was 8 mm at the outer surface of the pipe before welding. Tensile stresses on the

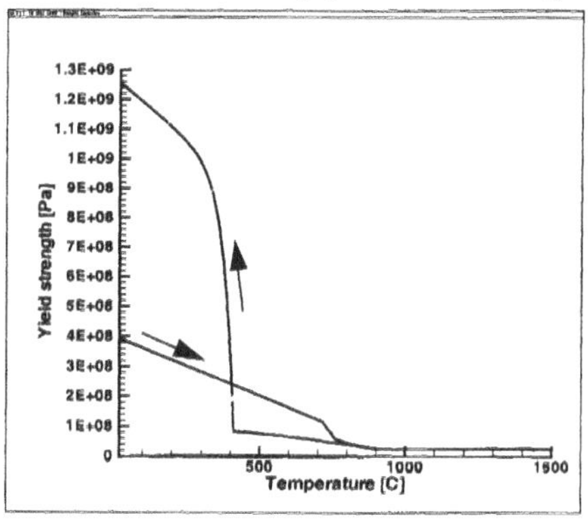

Figure 6. Computed virgin yield strength as a function of temperature. From [26].

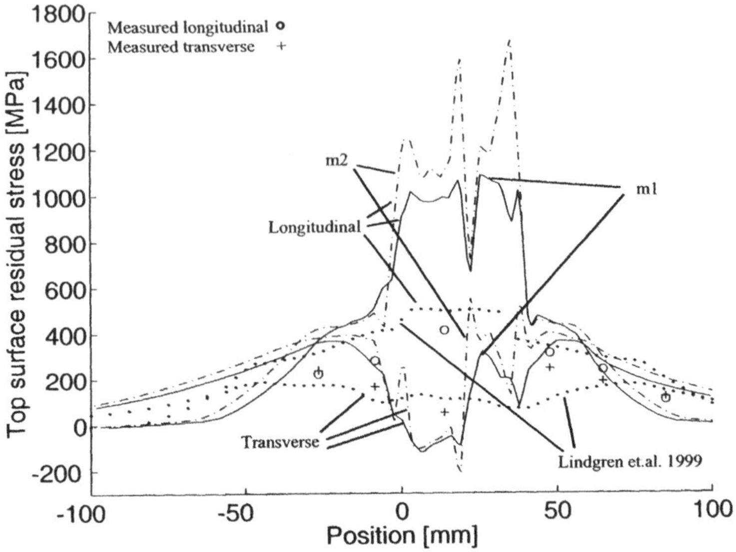

Figure 7. Calculated and measured residual stress at the top and bottom surfaces. The origin (x=0) is placed at the left side of the joint. From [26].The computational model with transformation induced plasticity is referred to as m1 and without transformation induced plasticity is referred to as m2

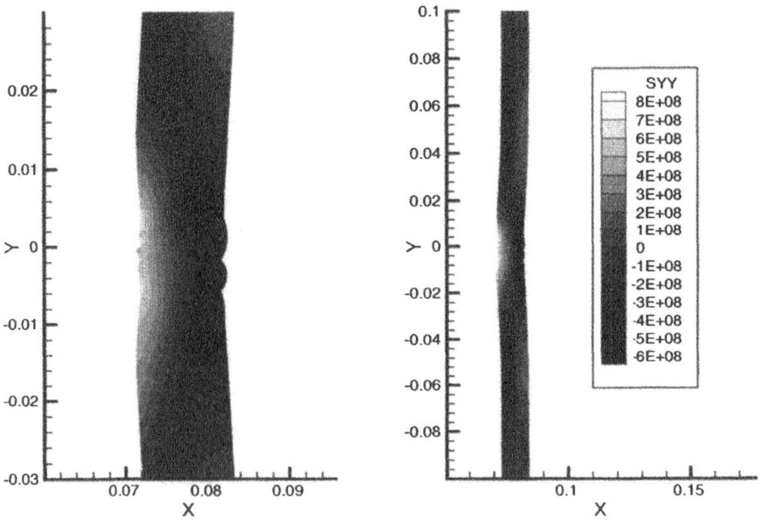

Figure 8. Residual axial stress for pipe welded by the U-seam procedure. Magnified view to the left. From [56].

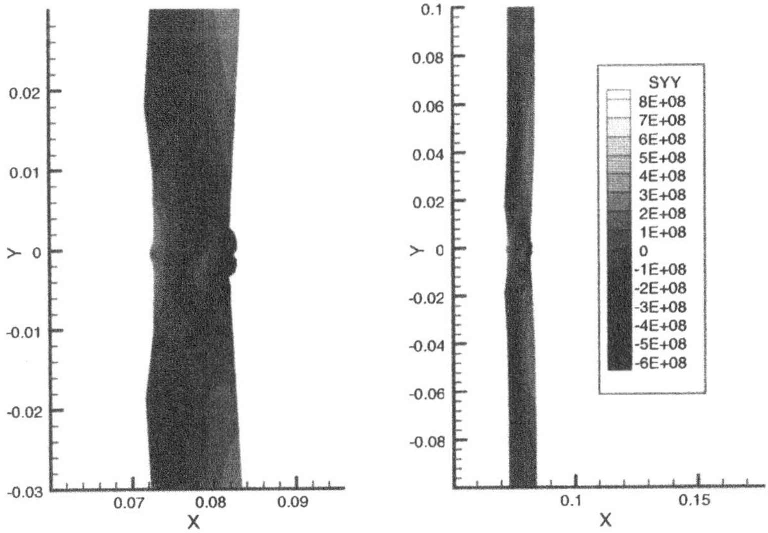

Figure 9. Residual axial stress for pipe welded by the Optigap procedure. Magnified view to the left. From [57].

inside of the pipe promote Internal Grain Stress Corrosion Cracking (IGSCC). Therefore, it is of interest to use a WPS which minimise this stress. The axial stress, which is of special interest for propagating cracks, is shown in Figures 8-9. It is found that the region with tensile stresses on the inside of the pipe is smaller and the stresses are lower.

7 Conclusions

There has been a very strong and successful development of the finite element method and it is expected that welding simulations can be performed in real time within the next decade. There are some challenges faced today in the field of welding simulations.

- Numerical techniques for modelling multipass welding must be improved. This includes methods for automatic creating the model of the added filer material.
- Microstructure models with mixture rules which makes it possible to account for the history dependency of the material properties are needed.
- More advanced constitutive models suitable for modelling material behaviour at higher temperatures.
- Models for the effect of residual stresses on life time. This includes models for ratchetting and shake-down effects.

However, the greatest obstacle is the lack of data for existing material models. Thus an increase in simulation capacity must be accompanied with an increased in experimental programmes supplying the needed data [58].

8 References

[1] Okumoto, Y., and Matsuzaki, S., Study of accuracy control of hull structure, *Journal of ship Production*, Vol 10, pp. 51, 1994.

[2]Tsai, C.L., Park, S.C., and Cheng, W.T., Welding distortion of a thin-plate panel structure, *Welding Journal,* Vol. 78, pp. 156s, 1999.

[3] Wang, G.S., To analyse fatigue crack growth in residual stress field at welded joints. *Proceedings of the Sixth International Fatigue Congress FATIGUE'96*, Elsevier Science, Oxford, England, pp 1219, 1996.

[4] Lopez, L., Martinez, Lin Peng, R., Blom, A.F., and Wang, G.S., Welding and TIG-dressing induced residual stresses relaxation and influence on fatigue strength of spectrum loaded weldments, Fatigue Design and Reliability, Ed:s G. Marquis and J. Solin, *ESIS Publication 23*, Elsevier Science, Oxford, England, pp. 117, 1999.

[5] Janosch, J.J., Debiez, S., Clerege, M., and Dang Van, K., Application of the local mechanical approach for optimizing the fatigue strength of fillet weld assemblies, *Proceedings of the IIW 50th Annual Assembly Conference on Performance of Dynamically Loaded Welded Structures*, Ed:s, Maddox, S.J., and Prager, M., Welding Research Council, New York, NY, USA, pp. 265, 1997.

[6] Henrysson, H.F., Abdulwahab, F., Josefson, B.L., and Fermer, M., Residual stresses in resistance spot welds: finite element simulations, x-ray experiments and influence on fatigue behaviour, *Welding in the World,* Vol. 43, pp. 55, 1999.

[7] Michaleris, R, Kirk, M., Mohr, W., and McGaughy, Incorporation of Residual Stress Effects into Fracture Assessments via the Finite Element Method, *Fatigue and Fracture Mechanics: 28th Volume, ASTM STP 1321*, Ed:s: J. H. Underwood, B. D. Macdonald, and M. R. Mitchell, American Society for Testing and Materials, pp. 499, 1997.

[8] Brust, F.W., Dong, P., and Zhang, J., Influence of residual stresses and weld repairs on pipe fracture, *Approximate Methods in the Design and Analysis of Pressure Vessels and Piping Components, ASME PVP-Vol. 347*, Ed. W.J. Bees, American Society of Mechanical Engineers, New York, NY, USA, pp. 173, 1997.

[9] Zhang, J., and Dong, P., Residual stresses in welded moment frames and implications on structural performance, to appear in *ASCE Journal of Structural Engineering*, 1999.

[10] Brickstad, B., and Josefson, B.L., A parametric study of residual stresses in multi-pass butt-welded stainless steel pipes, *International Journal of Pressure Vessels and Piping*, Vol. 75, pp. 11, 1998.

[11] Michaleris, P., and deBiccari, A., A predictive technique for buckling analysis of thin section panels due to welding, *Journal of Ship Production*, Vol. 12, pp. 269, 1996.

[12] Michaleris, P., and Sun, X., Finite element analysis of thermal tensioning techniques mitigating weld buckling distortion, *Welding Journal*, Vol. 76, pp. 451s-457s, 1997.

[13] Alfredsson, K.S., and Josefson, B.L., Harmonic response of a spot-welded box beam - influence of welding residuals stresses deformations, *Welding in the World,* Vol 30, pp 229-235, 1992.

[14] Josefson, B.L. and Lindgren, L.-E., Modelling of residual stresses, *The Fifth International Conference on Residual Stresses (ICRS-5)*, Ed:s Ericsson, T., Odén, M., and Andersson, A., Linköping University, Linköping, Sweden, pp. 1-11, 1997.

[15] Josefson, B.L., Prediction of residual stresses and distortions in welded structures. *ASME Journal of Offshore Mechanics and Arctic Engineering*, Vol 115, pp 52-57,1993.

[16] Radaj, D., *Heat Effects of Welding*, Springer Verlag, Berlin, Germany, 1992.

[17] Leggatt, R.H., Welding residual stresses, *The Fifth International Conference on Residual Stresses (ICRS-5)*, Ed:s Ericsson, T., Odén, M., and Andersson, A., Linköping University, Linköping, Sweden, pp. 12-25, 1997.

[18] *Handbook of Measurement of Residual Stresses*, Society for Experimental Mechanics, Ed. J. Lu, Fairmont Press, Liburn, GA, USA, 1996.

[19] Tsai, C.L., Modeling of welding-induced plasticity for distortion prediction, Document IIW-X/XV-RSDP-35-99, International Institute of Welding, 1999.

[20] Josefson, B.L., and Karlsson, C.T., FE-calculated stresses in a multi-pass butt-welded pipe – a simplified approach, *International Journal of Pressure Vessels and Piping*, Vol. 38, pp. 227-243, 1989.

[21] Josefson, B.L., Wikander, L., Hederstierna, J., and Johansson, F., Welding residual distortions in ring-stiffened pipes, *ASME Journal of Offshore Mechanics and Arctic Engineering*, Vol 118, pp 121-126, 1996.

[22] Wilkening, W.W., and Snow, J.L., Analysis of welding-induced residual stresses with the ADINA-system, *Computers & Structures*, Vol. 47, pp. 767-786, 1993.

[23] Sun, X., Dong, P., and Kimchi, M., The coupled electrical-thermal-mechanical process associated with Aluminium RSW, Body Assembly and Manufacturing, *Proceedings of the International Body Engineering Conference (IBEC)*, Stuttgart, Germany, Vol. 30, pp. 42-48, 1997.

[24] Andersson, B.A.B., Thermal stresses in a submerged-arc welded joint considering phase transformations, *ASME J. of Engineering Materials and Technology*, Vol. 100, pp. 356, 1978.

[25] Bergheau, J.M., and Leblond, J.B., Coupling between heat flow, metallurgy and stress-strain computation in steels: the approach developed in the computer code Sysweld for welding and quenching, *Proc. Modeling of Casting, Welding and Advanced Solidification Processes V*, The Minerals, Metals & Materials Society, pp. 203, 1991.

[26] Börjesson, L., and Lindgren, L-E., Simulation of multipass welding with simultaneous computation of material properties, submitted for publication.

[27] Lewis, R.W., Morgan, K., Thomas, H.R., Seetharamu, K.N., *The Finite Element Method in Heat Transfer Analysis*, Wiley, 1996.

[28] Rolph, W. D., and Bathe, K-J., An efficient algorithm for analysis of non linear heat transfer with phase changes, *Int. J Numerical Methods in Engineering*, Vol. 18, pp. 119, 1982.

[29] Soltani, B., Mattiasson, K., and Samuelsson, A., Comparison between solid and flow approaches in blade forging, Numerical Methods in Engineering'96, *Proceedings of Second ECCOMAS Conference on Numerical Methods in Engineering*, Sep. 1996, Paris, France.

[30] Atteridge, D.G., Becker, M., Meekisho, L., Owusu-Nelson, K., Thamasebi, B., and Zhang, L., 3D Eularian modeling of electroslag welding, *Proc. of the 5th Int. Conf. on Trends in Welding Research*, Pine Mountain, USA, June, pp. 943, 1998.

[31] Goldak, J.A., and Gu., M., Computational weld mechanics of the steady state, *Proc. of the 2nd Int. Conf. on Mathematical Modelling of Weld Phenomena* (ed. Cerjak), Material Modelling Series (ed. Bhadeshia), The Institute of Materials, pp. 207, 1995.

[32] Belytschko, T., and Hughes, T.J.R. (eds), *Computational Methods for Transient Analysis*, Volume 1 in Computational Methods in Mechanics, North-Holland, 1983.

[33] Mahin, K.W., MacEwen, S., Winters, W., Mason, W., Kanouf, M., and Fuchs, E.A., Evaluation of residual stress distributions in a traveling GTA weld using finite element

and experimental techniques, *Proc. Modeling of Casting and Welding Processes IV, The Minerals*, Metals & Materials Society, pp. 339, 1988.

[34] Zienkiewicz, O.C., and Taylor, R.L., *The Finite Element Method*, McGraw-Hill, 1989.

[35] Crisfield, M.A, *Non-linear Finite Element Analysis of Solids and Structures, Vol. 1 Essentials*, J Wiley & Sons, 1991.

[36] Brown, S.B., Christie, K., and Song, H., Three dimensional finite element modeling of welded structures, *Mechanical Effects of Welding*, IUTAM Symposium Luleå/Sweden 1992 (eds L. Karlsson, L-E. Lindgren and M. Jonsson), pp. 181, Springer-Verlag, 1992.

[37] Bergheau, J.M., Pont, D., and Leblond, J.B., Three dimensional simulation of a laser surface treatment through steady state computation in the heat source's comoving frame, *Mechanical Effects of Welding*, IUTAM Symposium Luleå/Sweden 1992 (eds Karlsson, L., Lindgren, L-E. and Jonsson, M.), pp. 85, Springer-Verlag, 1992.

[38] Lindgren, L-E., Häggblad, H-Å., McDill, J.M.J., and Oddy, A.S., Automatic remeshing for three-dimensional finite element simulation of welding, *Computer Methods in Applied Mechanics and Engineering*, Vol. 43, No. 4, pp. 587, 1998.

[39] Runnemalm, H., and Hyun, J., Three Dimensional Welding Analysis using an Adaptive Mesh Scheme, accepted for publication in *Computer Methods in Applied Mechanics and Engineering*.

[40] Simo, J.C., A framework for finite strain elastoplasticity based on maximum plastic dissipation and the multiplicative decomposition, *Computer Methods in Applied Mechanics and Engineering*, Vol. 66, pp. 199, 1988.

[41] Cuitino, A., and Ortiz, M., A material-independent method for extending stress update algorithms from small strain plasticity to finite plasticity with multiplicative kinematics. *Engineering Computations*, Vol. 9, pp. 437, 1992.

[42] Lee, E.H., Elastic-plastic deformation at finite strain, *ASME Journal of Applied Mechanics*, Vol. 36, pp. 1, 1969.

[43] Simo, J.C., and Pister, K.S., Remarks on rate constitutive equations for finite deformation problems: Computational implications, *Computer Methods in Applied Mechanics and Engineering*, Vol. 46, pp. 201, 1984.

[44] Crisfield, M.A., *Non-Linear Finite Element Analysis of Solids and Structures, Vol. 2 Advanced topics*, J Wiley & Sons, 1997.

[45] Hughes, T.J.R., Numerical Implementation of Constitutive Models: Rate-independent Deviatoric Plasticicty, *Workshop on the Theoretical Foundation for Large-Scale Computations of Nonlinear Material Behavior*, Northwestern University, Evanston, Illinois, 24-26 Oct 1983.

[46] Lindgren, L-E., A welding simulation revisited, *3rd Int. Conf. on Thermal Stresses*, Cracow, Poland, June, 1999.

[47] Lemaitre, J., and Chaboche, J-L., *Mechanics of Solid Materials*, Cambridge University Press, 1990.

[48] Ekh, M., Johansson, A, Thorberntsson, H, and Josefson, B.L., 2000, Models for cyclic ratchetting plasticity - Integration and Calibration, *ASME Journal of Engineering Materials and Technology*, Vol. 122, pp. 49-55.

[49] Kojic, M., and Bathe, K-J., The "effective-stress-function" algorithm for thermo-elasto-plasticity and creep, *International Journal for Numerical Methods in Engineering*, Vol. 24, pp. 1509, 1987.

[50] Leblond, J.B., A theoretical and numerical approach to the plastic behaviour of steels during phase transformations - I. Derivation of general relations and -II. Study of classical plasticity for ideal plastic phases, *Journal of the Mechanics and Physics of Solids*, Vol. 34, No 4, pp. 395, 1986.

[51] Svoboda, A., Lindgren, L-E., and Oddy, A.S., The effective stress function algorithm for pressure dependent plasticity applied to hot isostatic pressing, in *Computational Modelling of Hot Isostatic Pressing*, Doctoral thesis LTU-DT--1997/7, Luleå University of Technology, Sweden, 1997.

[52] Lindgren, L-E., Finite element modelling and simulation of welding, submitted for publication.

[53] Sheng, I.C., and Chen, Y., Modeling welding by surface heating, *ASME J. of Engineering Materials and Technology*, Vol. 114, No. 4, pp. 439, 1992.

[54] Brust, F.W., Dong, P., and Zhang, J., A constitutive model for welding process simulation using finite element methods, *Advances in Computational Engineering Science*, S.N. Atluri and G. Yagawa (eds), pp. 51., 1997.

[55] Tekwriwal, P., and Mazumder, J., Transient and residual thermal strain-stress analysis of GMAW, *ASME J. of Engineering Materials and Technology*, Vol. 113, pp. 336, 1991.

[56] Lindgren, L-E., Runnemalm, H., and Näsström, M.O., Numerical and experimental investigation of multipass welding of a thick plate, *International Journal for Numerical Methods in Engineering*, Vol. 44, No. 9, pp. 1301-1316, 1999.

[57] Hedblom, E., Lindgren, L-E., Nordgren, G., and Gillander, L., Welding procedures with reduced residual stresses for components in nuclear power plants, *15th Int. Conf. on Structural Mechanics in Reactor Technology (SMiRT-15)*, Seoul, Korea, Aug., 1999.

[58] Lindgren, L-E., The Use of Simulations and the Need of Experiments in Material Processing, Keynote at the *sixth Cairo International MDP Conference*, pp. 149, Cairo, 1996.

S708/002/2000

Modelling of residual stresses at girth welds in pipes

E OLDEN and **R LEGGATT**
The Welding Institute, Cambridge, UK

SYNOPSIS

Many organisations are using numerical modelling to calculate the residual stresses due to welding, but how consistent are the results? This paper describes TWI's contribution to a Round Robin on weld modelling organised by the International Institute of Welding (IIW). The first phase of the Round Robin addresses distortion and residual stresses at girth welds in pipes.

1. INTRODUCTION

In 1996 the International Institute of Welding (IIW) formed a working group to study residual stresses and distortion prediction in welded structures. It was decided that the working group would focus on the following objectives:

- Developing a compendium of residual stress and distortion solutions
- Conducting a Round Robin to validate and benchmark residual stress and distortion prediction codes
- Developing improved modelling approaches

A proposal was made for the Round Robin to be conducted in two phases. Phase 1 commenced in February 1999 and is due to finish at the end of this year. This paper describes the protocol for the first phase, TWI's contribution to the work, and an initial analysis of the results.

In Phase 1 participants were required to predict the axial and hoop (relative to the pipe) residual stress distributions for two cases of pipe girth welds. Phase 2 is likely to start in 2000 and will involve residual stress and distortion prediction for a stiffened panel and a butt weld in an austenitic stainless steel plate. The latter case will also include an experimental investigation of the residual stresses.

2. PROTOCOL FOR PHASE 1

The pipe weld preparation that was modelled is shown in Figure 1. Axisymmetric conditions were used to model a single-V girth butt weld in a pipe of wall thickness 19 mm and inner radius of 190 mm. The weld joint had an included angle of 60°, a root gap of 3.18 mm and 3.18 mm land. Figure 2 illustrates the welding sequence for Cases 1 & 2. Case 1 is a three-pass girth weld with a relatively high heat input per pass and Case 2 is an eight-pass girth weld with a relatively low heat input per pass.

In order to ensure a consistent approach was adopted, participants in the Round Robin were provided with pipe and weld geometry, finite element meshes, and material property data as a function of temperature. The material properties are listed in Table 1. They were not necessarily typical of a carbon steel. Also provided were the welding procedure in terms of the travel speed and net power for the torch for each pass in the welds (Tables 2a & 2b). Because "net" heat input was defined it was not necessary for assumptions to be made regarding the arc efficiency.

3. TWI'S CONTRIBUTION TO PHASE 1.

The commercial finite element package, ABAQUS (Version 5.8) was used to perform the analyses. An uncoupled thermo-mechanical approach was adopted. Initially a thermal distribution was obtained following the addition of each weld pass. These thermal strains were then used to perform a mechanical analysis to determine the pipe distortion and distribution of residual stresses.

3.1. Thermal Analysis

Heat was added to the model as a body flux applied as each pass was made. An amplitude curve was used to define the addition of heat with time and is illustrated in Figure 3. Following the addition of heat from each pass, the model was allowed to cool for the approximate time required for the torch to pass around the pipe circumference, or until the maximum temperature was less than 200°C, whichever took the longer. Therefore the interpass temperature was limited to 200°C in the TWI analysis. The thermal analysis was stopped after the cooling of each pass to determine the initial temperature conditions to be used when subsequent passes were added. Heat redistribution occurred by conduction, latent heat, and convection and radiation from the pipe walls.

3.2. Mechanical Analysis

In the mechanical analysis, the thermal strains obtained from the thermal analysis were used to determine residual stresses in the pipe. The pipe ends were free to contract during the welding simulation. Elements within each pass were added in a strain free condition. They were only activated in the model after the thermal analysis for that pass showed that the material had cooled to at least 1500°C. TWI assumed pure isotropic hardening for the model.

4. Results from TWI Analyses

4.1. Temperature Results

The variation of temperature with time was plotted for five points on the pipe for both cases (see figs. 4 & 5). Three of the points were at the inner surface at 12.7 mm, 19 mm and 25 mm from the weld centreline, and two on the outer diameter at 19 mm and 25 mm from the weld centreline.

4.2. Residual Stresses

Residual stress values were determined after specific stages in the welding sequence along two lines through the pipe wall thickness (indicated on Figure 1). The graph in Figure 6 shows the predicted residual stresses at location 1 for both Case 1 and 2 after the final weld pass. The stresses in the hoop direction are parallel to the weld, and the axial stresses act perpendicular to the weld. Stress values have been normalised by the room temperature yield strength for the material.

4.3. Distortion Results

Figure 7 shows the results for radial distortion for the two cases.

5. Discussion of the Results from TWI

The temperature profiles in figures 4 & 5 show that, as would be expected, the maximum temperatures are reduced for Case 2 (the 8 pass weld) in comparison to Case 1.

The residual stresses through the weld thickness at the centre line of the weld (figure 6) show not only different magnitudes of stress for the different cases, but also different distributions. For Case 1 both hoop and axial stresses were highest near the inner diameter and decreased through the thickness. However, for Case 2, the stresses were approximately sinusoidal and were at a maximum in the hoop direction (parallel to the weld) at approximately the mid-thickness.

Figure 7 indicates that the inwards radial displacement at the root of the weld is greater for Case 2 than for Case 1.

6. Preliminary Comparison of Results from Participants

In July 1999 preliminary results were presented at the IIW Annual Assembly from 6 of the organisations. As seventeen organisations had indicated that they were intending to participate in the Round Robin, only a few sets of results were shown to avoid influencing the remainder of the organisations involved. The results that have been released by EWI are shown in Figures 8 and 9. The results were analysed by EWI who noted the following general trends:

- Results showed a higher level of variability than had been expected.
- In general there was better agreement between the results for Case 1 than for Case 2.
- The poor agreement in peak temperature predictions as shown in Figure 8 may explain variations in residual stresses
- The predicted transverse residual stress distribution for Case 1 showed high levels of variability, particularly in the mid-wall region, as shown in Figure 9.

The organisers of the Round Robin have identified the following parameters as possibly being responsible for the observed trends and have decided they should be analysed in more detail:

- The material hardening model, kinematic or isotropic
- The length of the finite element model
- How the model was restrained during welding
- Heat flux
- Interpass temperature

7. Conclusions

A Round Robin exercise to predict residual stresses and distortion at girth welds in pipes for two cases (Case 1 - 3 pass weld, Case 2 - 8 pass weld) has been described, and the contribution of TWI and preliminary results from other organisations presented. The results of TWI have indicated:

- Lower peak temperatures were recorded when 8 passes were used to generate the weld rather than 3 passes in the same wall thickness
- Predicted residual stress distributions vary for the two cases although peak stress magnitudes are not very different along the weld centreline
- The inwards radial displacements at the root of the weld are lower for Case 1 than for Case 2.

An initial analysis of the results from other organisations has shown that there is a larger than expected variability in the results from different analysts. Possible factors for these discrepancies have been identified and will be looked at more closely. Eventually it is hoped that the Round Robin will result not only in a benchmark for residual stress and distortion modelling, but also provide guidelines for this type of analysis.

Temp. (C)	Yield stress (MPa)	Poisson's ratio	Young's modulus (GPa)	Plastic modulus (GPa)	Convection coeff. (W/(m^2 C))	Coeff. thermal expansion (*10^-6) (1/C)	Specific heat (J/Kg C)	Conducti on coeff. (W/m C)	UTS at 0.02 plastic strain(MPa)
0	250	0.3	200	10	4	10	300	65	450
100	245	0.32	198	9.5	5	10	301	60	435
200	240	0.34	196	9	6	10	304	55	420
300	235	0.36	194	8.5	6.75	10.8	309	50	405
400	230	0.38	192	8	7.5	11.5	317	45	390
500	225	0.4	190	7.5	8.25	12.3	328	40	375
600	205	0.42	180	6.5	9	13	343	35	335
700	185	0.44	165	5.5	9.75	13.8	361	30	295
800	165	0.46	150	4.5	10.5	14.5	384	25	255
900	145	0.48	135	3.5	11.25	15.3	411	20	215
1000	125	0.499	120	2.5	11.75	16	443	20	175
1100	105	0.499	105	1.5	12.15	16	479	20	135
1200	85	0.499	90	1	12.55	16	521	20	105
1250								20	
1300	65	0.499	75	0.75	12.85	16		45	80
1350								70	
1400	45	0.499	60	0.5	13.15	16	526	95	55
1450								120	
1500	40	0.499	45	0.4	13.55	16		145	48
1600	35	0.499	44	0.3	13.85	16	531	145	41
1700	30	0.499	43	0.3	14.05	16		145	36
1800	25	0.499	42	0.3	14.25	16	536	145	31
1900	20	0.499	41	0.3	14.45	16		145	26
2000	15	0.499	40	0.3	14.6	16	541	145	21
2100	15	0.499	40	0.3	14.725	16		145	21
2200	15	0.499	40	0.3	14.85	16	546	145	21
2300	15	0.499	40	0.3	14.975	16		145	21
2400	15	0.499	40	0.3	15.075	16	551	145	21
2500	15	0.499	40	0.3	15.15	16		145	21
2600	15	0.499	40	0.3	15.15	16	556	145	21
2700	15	0.499	40	0.3	15.15	16		145	21
2800	15	0.499	40	0.3	15.15	16	561	145	21
2900	15	0.499	40	0.3	15.15	16		145	21
3000	15	0.499	40	0.3	15.15	16	566	145	21
Latent Heat (J/Kg)	Solidus (C)	Liquidus(C)							
2.47E+05	1455	1554							

Table 1. Summary of material properties used in the finite element analyses

Pass	Travel speed (in./min)	Net power (watts)
1	9.5	1600
2	7.51	5300
3	7.51	8000

Table 2a. Heat input parameters for Case 1.

Pass	Travel speed (in./min)	Net power (watts)
1	9.5	1600
2	9.5	4300
3	7.51	1400
4	7.51	1200
5	4.8	1900
6	4.8	1900
7	4.8	1900
8	4.8	2400

Table 2b. Heat input parameters for Case 2.

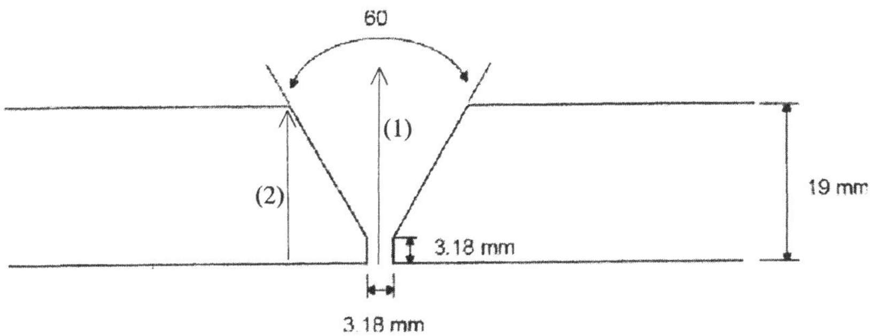

Figure 1. Weld joint design for girth weld in cylinder

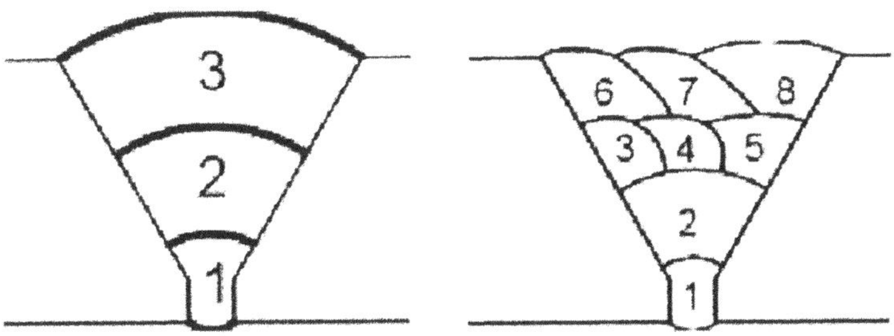

Figure 2. Welding sequence for Case 1 (3 passes) and Case 2 (8 passes)

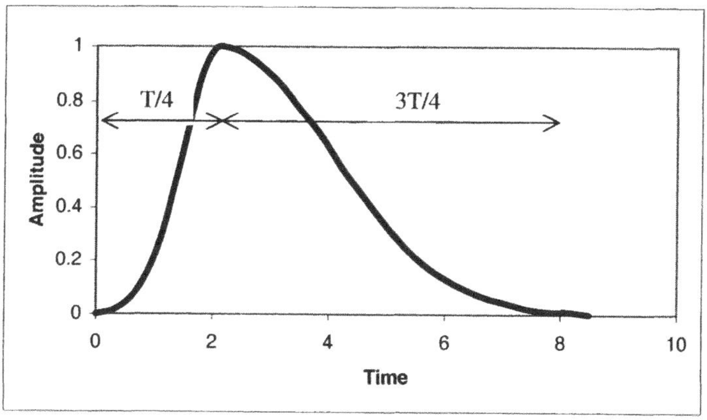

Figure 3. Amplitude curve used to define heat input (T = total time).

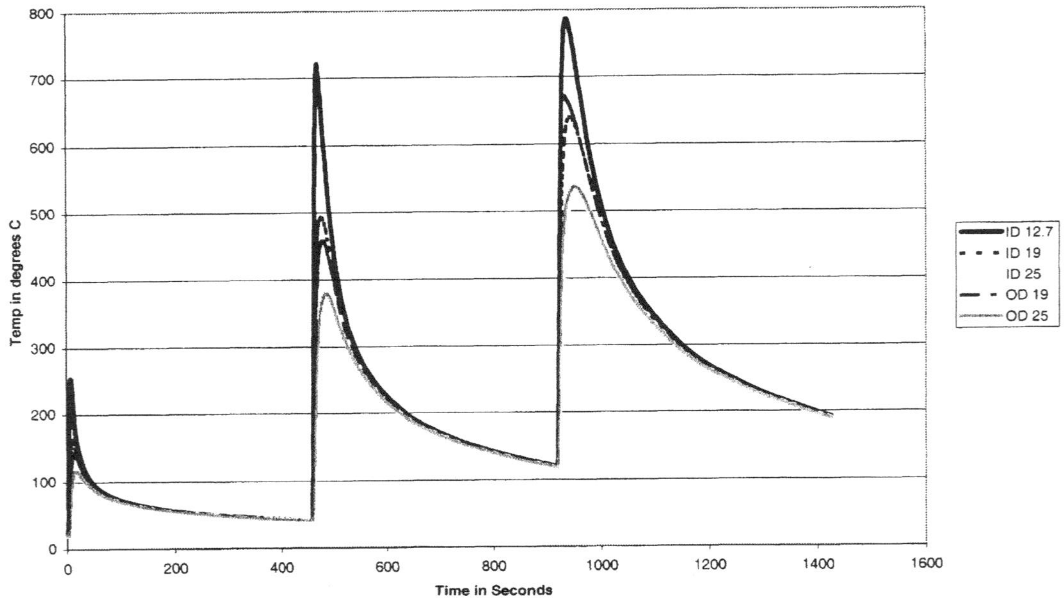

Figure 4. Temperature against time for Case 1 (3 Passes)

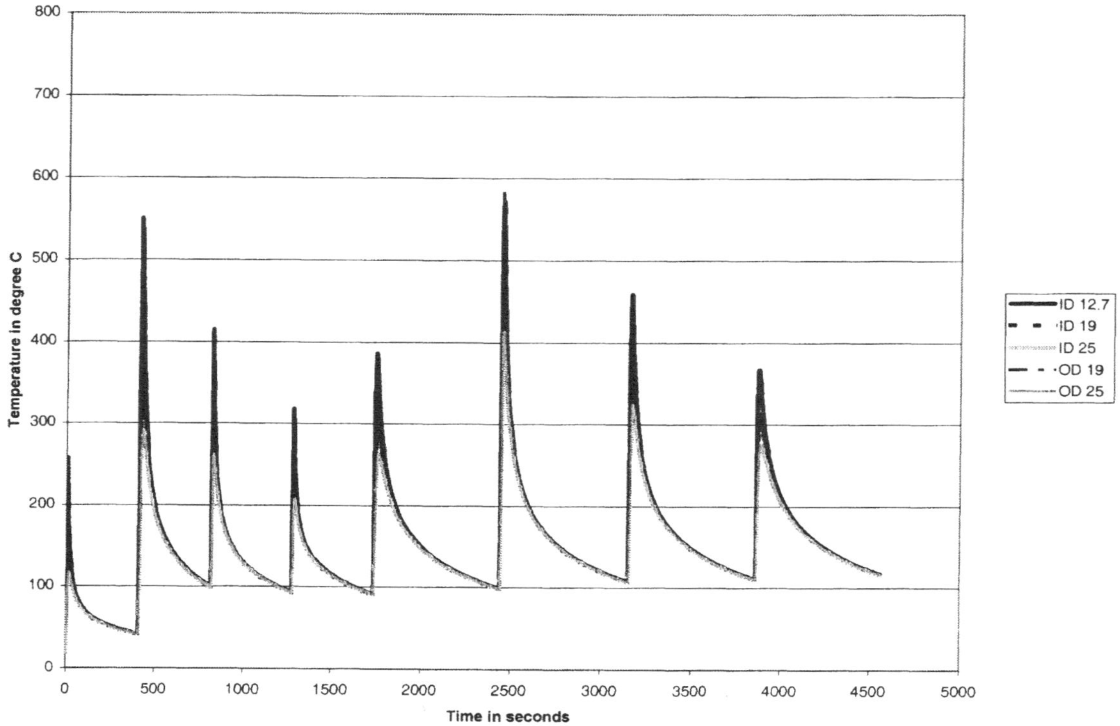

Figure 5. Temperature against time for Case 2 (8 Passes)

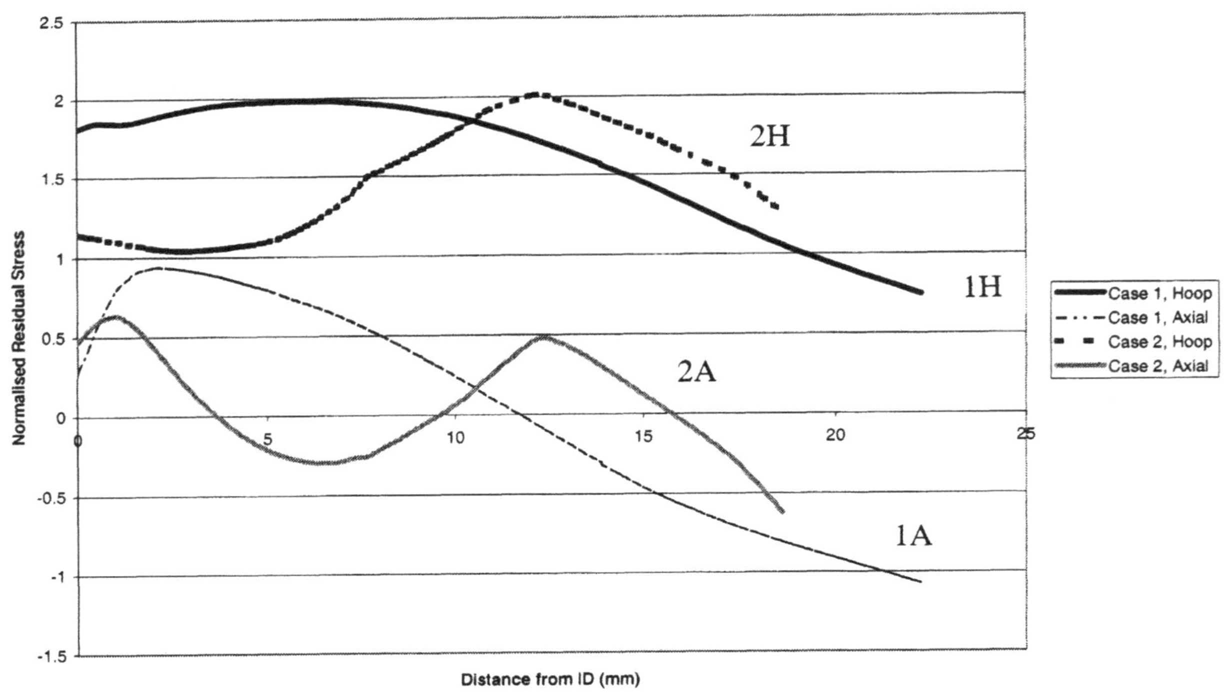

Figure 6. Predicted residual stress values for the two cases.

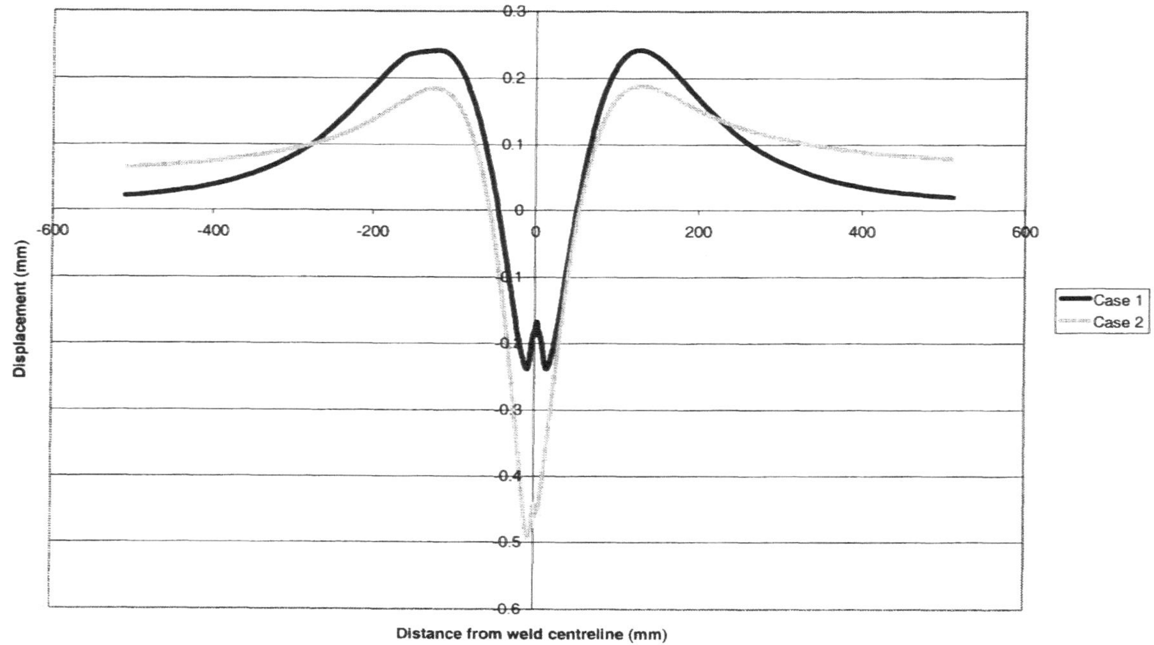

Figure 7. Results of radial distortion for the two cases.

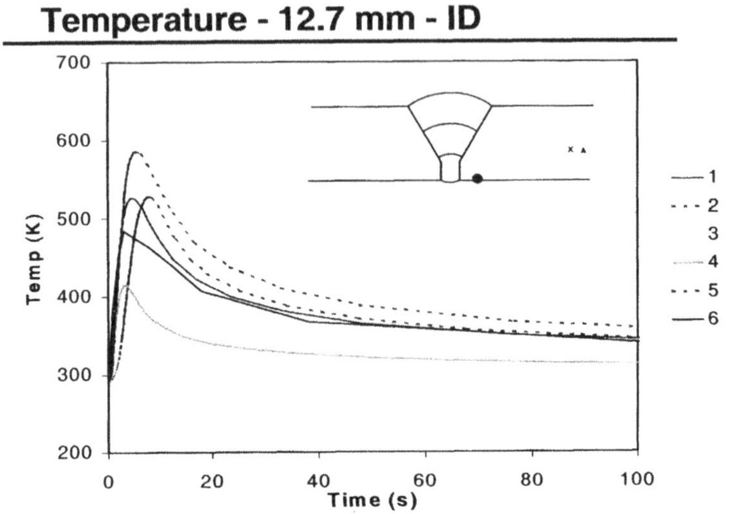

Figure 8. Results from 6 organisations for temperature variation with time at 12.7 mm from the weld centreline on the inner surface of the pipe after the final pass for Case 1.

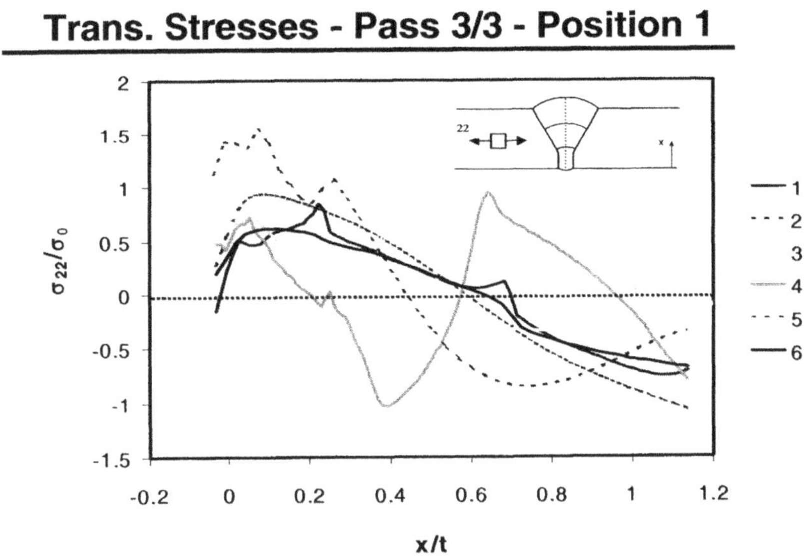

Figure 9. Results from 6 organisations for transverse stresses through the thickness at the weld centreline after the final pass for Case 1.

S708/003/2000

Nickel-base superalloy welding – models for residual stress, distortion, and weldability

S M ROBERTS, O HUNZIKER, D DYE, H J STONE, and **R C REED**
Rolls-Royce University Technology Centre, University of Cambridge, UK

SYNOPSIS

In this paper we present finite element based models for the welding of nickel-base superalloys common to many aeroengine components. Two welding processes are modelled; Electron Beam Welding (EBW) commonly employed in the joining of turbine discs and Tungsten Inert Gas (TIG) welding for combuster liner and casings applications. Both models share the common basis of a thermal model sequentially coupled to an elasto-plastic mechanical model. For the TIG welding process a microstructural model for the diffusion of the alloying elements has also been added to provide some measure of 'weldability'. The approach developed enables the weld engineer to predict the appropriate welding conditions to minimise weld induced distortion and the likelihood of cracking due to solidification or liquation and provides valuable residual stress input for component lifing calculations.

In order to provide model validation Waspaloy plates have undergone bead-on-plate EBW, whilst IN718 sheet material has been TIG welded. In-situ thermocouples have provided transient temperature profile verification and post weld distortion has been characterised by needle probe analysis. Residual stress measurements have been performed using hole drilling and diffraction techniques. Electron Probe Micro Analysis (EPMA) has been used for the evaluation of segregation.

1. INTRODUCTION

Welding is now a common method of fabrication in gas turbines. However, determination of the appropriate parameters to successfully weld a component requires significant empirical knowledge and practical experience. Today, with modern computing resources, numerical based approaches, such as the Finite Element Method (FEM), are capable of predicting both distortion and residual stress 1,2,3 and may even be coupled with rules defining the onset of cracking4.

EBW is commonly applied in the welding of both nickel base and titanium alloy disc components. The significant amount of highly focussed power permits welds of high aspect ratio and hence minimal distortion. At the same time conducting the weld in a vacuum chamber reduces the risk of weld contamination.

TIG welding is used in both the manufacture and repair of modern jet engine combustors. Again the technique is well developed, but with a requirement for ever hotter engine temperatures, more and more exotic nickel-base alloys are being requested. Both solidification cracking and liquation cracking may result from the use these alloys.

The approach taken in our group has been to develop appropriate heat sources to match the deposition of heat into the material. A thermal analysis is then performed to calculate the spatial and temporal evolution of temperature. This temperature evolution then drives a mechanical analysis via the coefficient of thermal expansion. For the cracking model the thermal model also acts as an input for the diffusion calculations.

Significant effort has been expended on the validation techniques necessary to produce reliable, accurate models. Thermocouples have been employed whenever possible to capture the thermal evolution. Neutron diffraction, X-ray diffraction and hole drilling methods have been used to measure the post weld residual stress. Distortion has been carefully characterised through the use of needle probe coordinate measurement devices. In order to identify the degree of microsegregation during welding EPMA analysis has also been performed. To understand the mechanical effect of the weld all test welds have been performed as bead-on-plate, with the plate unconstrained. A typical IN718 TIG weld is shown in Fig1., note the characteristic 'butterfly' hinging about the weld line and the longitudinal camber bend.

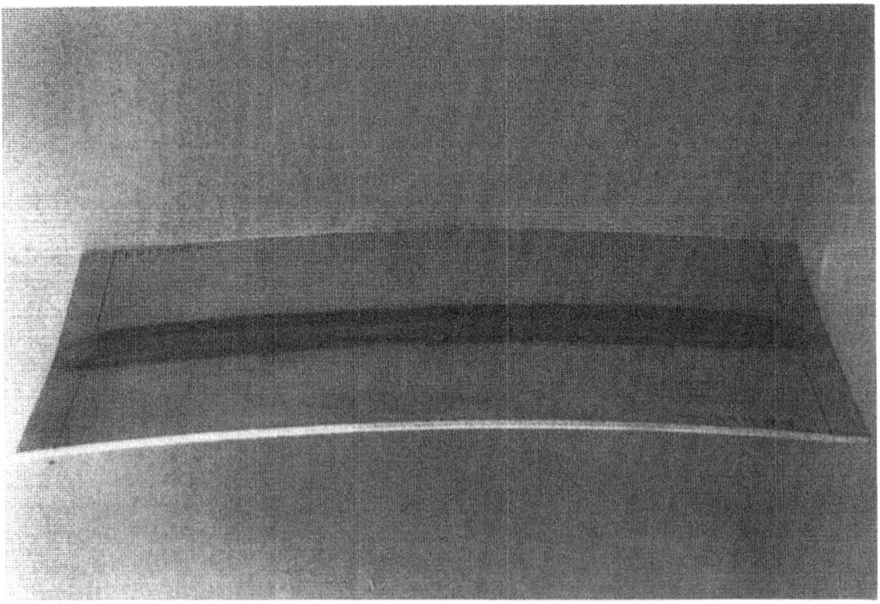

Fig 1. IN718 plate post weld, showing both butterfly and camber distortion

2. THERMAL ANALYSES

Two heat source terms have been developed for the two different welding processes. They are chosen so as to accurately represent the deposition of heat into the material being welded. At the same time the author's have found that computation times can be minimised through attempting to use the simplest adequate source. Care is also taken to ensure that sufficient discretisation is applied to enable the heat source to be accurately represented, in order to minimise calculation time remeshing techniques have been developed which are described elsewhere [1,5,6].

2.1 TIG Thermal Model

A surface disc source has been chosen for the IN718 TIG welding . The surface flux (q) for a weld of velocity v (x = x-vt) is given by

$$q\{x,y,z\} = hQ0 / pa2 \qquad \text{where } x2 + y2 < a2$$

where Q0 is the nominal power applied (~550W). h is the weld efficiency a measure of the amount of the power applied taken up by the material being welded, for TIG welding of IN718 this has been found to be around 50% to 60%. Typically a is of the order of the weld pool half width, in this case approximately 4mm. In order to determine a and h a series of steady state analyses are performed in the Eulerian reference frame. Once a good fit to the measured fusion boundary has been achieved, as shown in Fig 2., a full transient analysis may be performed. Such a full transient description is unnecessary for the microstructural model, and a semi-analytical model based on Rosenthal[7] or the steady state solution could be used.

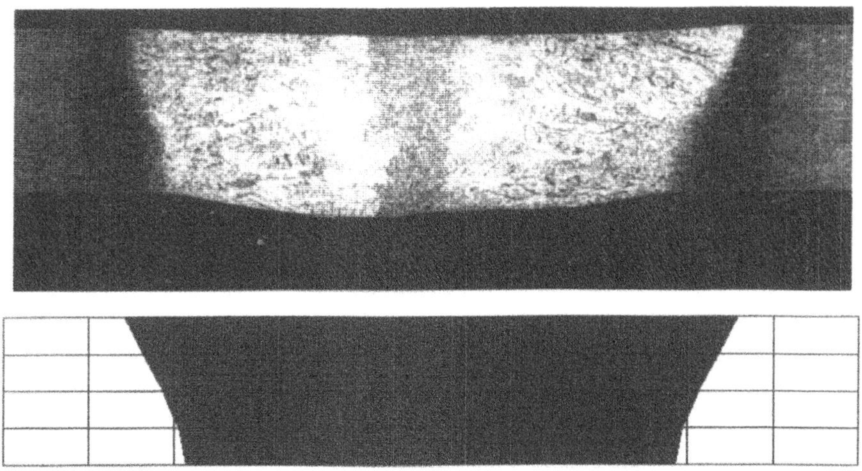

Fig 2. Comparison of fusion boundary profile and steady state melt isotherm – TIG weld in IN718

2.2 EBW Thermal Model

For the EBW model a more complex heat source description is necessary which represents both the stirring action of the molten metal and the digging action of the key hole. Again the stirring is well represented by a surface disc whilst a volumetric conic source has been found to replicate the deep penetration of the key hole. The resultant heat source is as follows :

$$q\{x,y,z\} = \frac{rhQ0}{pa2} \quad \text{where } x2 + y2 < a2$$

$$+ \quad \frac{(1\text{-}r)hQ0}{pd(1/3\ m2d2 + mdb + b2)} \quad \text{where } x2 + y2 < b2$$

Again Q0 is the nominal power applied (~2000W - 3500W dependent on thickness to be welded). The efficiency h for electron beam welding is greater than for TIG and can vary between approximately 80% for a focussed, penetrating weld to 98% for a heavily defocussed, non-penetrating weld. r is a partitioning coefficient whilst a,b and m are again fitting parameters which enable the user to fit the melt isotherm to the fusion boundary profile, d is the penetration depth of the weld. Two examples for welds in 9mm thick Waspaloy are shown in Fig3.

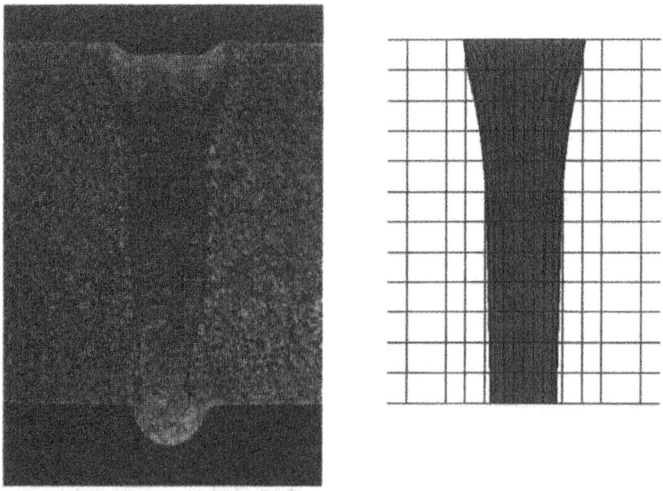

Fig 3a. Comparison of fusion boundary profile and steady state melt isotherm – EBW weld in Waspaloy – focussed condition

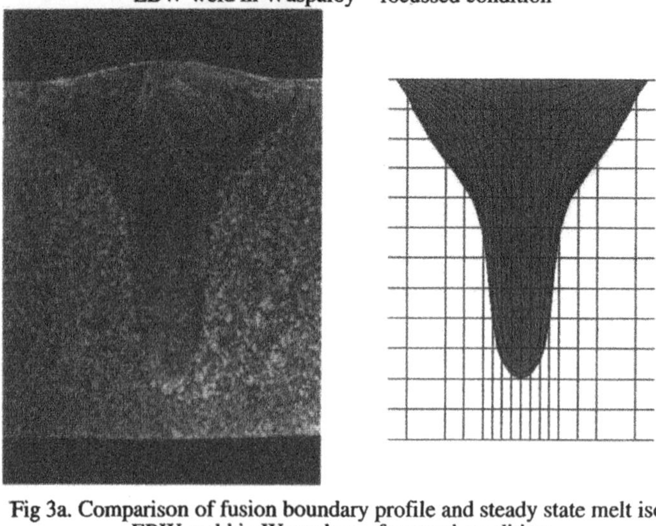

Fig 3a. Comparison of fusion boundary profile and steady state melt isotherm – EBW weld in Waspaloy – focussed condition

2.3 Thermocouple validation

Measurements of the thermal cycle in the heat affected zone have been taken using thermocouples soldered to the plate/sheet surface. As can be seen in Fig 4. satisfactory agreement is obtained between the measured and the predicted profiles. The upper and lower bound FEM solutions represent distances from the weld equivalent to the nearest and furthest part of the solder joint on a given thermocouple.

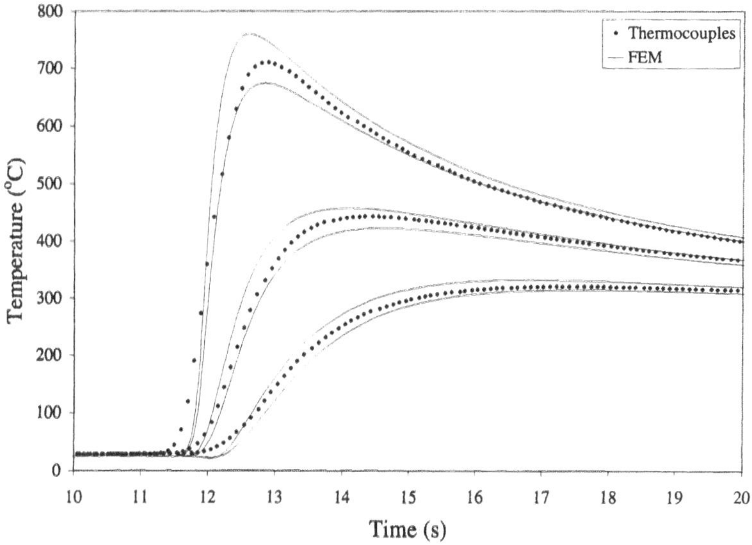

Fig 4. Comparison of measured thermocouple temperatures and predicted values – EBW plate

3. MECHANICAL ANALYSIS

Elasto-plastic 3-D finite element transient analysis is performed following methodologies described by the author's elsewhere.1,5,6. Care is taken to ensure strain compatibility between the thermal strains resulting from the thermal analysis and the strains within the mechanical field8. Typically the thermal model will be linear and the resultant thermal strain likewise. The strains in the mechanical analysis represent the first derivative of the displacement field. Choosing a quadratic element formulation results in appropriate compatibility, it also minimises the likelihood of the so-called shear locking associated with linear elements which are unable to reproduce pure bending behaviour9. By introducing reduced integration the volumetric locking which makes fully integrated elements behave in an overly stiff manner, due to the imcompressible plasticity criterion, may also be avoided.

3.1 Distortion Validation

Distortion in welded plates consists of two principal modes

- u a butterfly fold about the weld axis
- u a camber along the length in the weld direction,

Post-weld distortion measurements performed using a Leitz needle probe coordinate device confirms this. In Fig 5 the distortion out of the plane of a TIG welded sheet is compared to the numerical prediction.

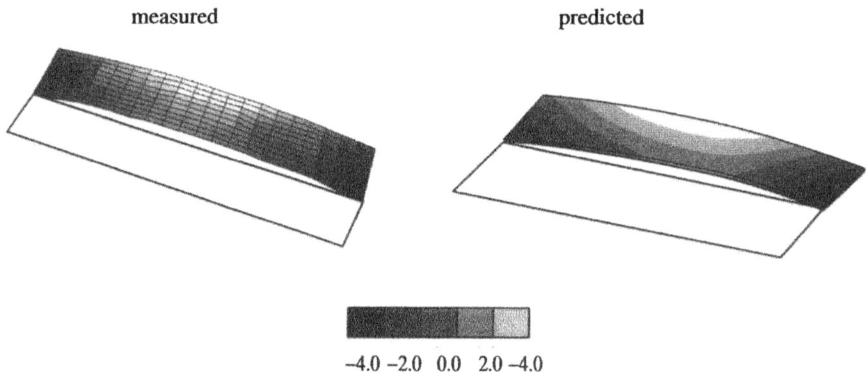

Fig 5. Distortion (mm) comparison for IN718 TIG weld

3.2 Residual Stress Validation

Incremental hole drilling, Xray diffraction and neutron diffraction have been used to characterise the residual stress state in our welded plates10. Details of the methods employed and their inherent strengths and weaknesses are given by Stone et al11. In brief, however both diffraction techniques make use of the structure of the material to act as an internal strain gauge. On deformation the lattice spacing is changed in proportion to the strain. By measuring the lattice spacing via the diffraction angle and comparing it to the undeformed state a measure of the strain is achieved. This may then be converted to residual stress in the standard way. X-rays suffer from poor penetration into Ni-base alloys and hence can only give a measure of surface stresses, whilst more expensive neutrons can penetrate the plates considered and give finer resolution. The hole drilling technique relies on measuring the relaxation around a hole drilled in stressed material via a strain gauge rosette. Inherently large sampling areas and the destructive nature of the technique may outweigh its comparative cheapness. Fig 6 shows predicted longitudinal residual stress compared to those measured via all techniques for an EBW plate.

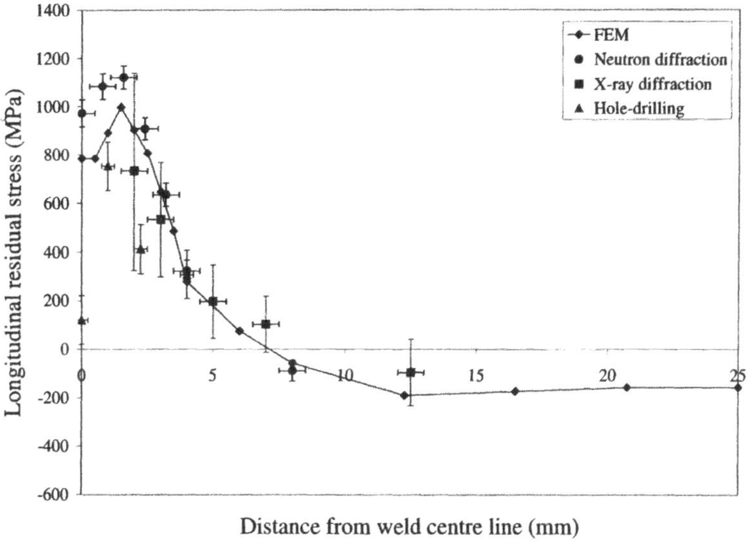

Fig 6. Comparison of longitudinal residual stress from EBW of Waspaloy

4. SOLIDIFICATION CRACKING MODEL

The methods we have developed to predict cracking due to welding are fairly unsophisticated but do appear to give the correct trends. The Scheil12,13 approach to solute redistribution during solidification is applied with the assumptions of "infinite" mixing in the liquid and no diffusion allowed within the solid. Such assumptions lead to an over-prediction of the amount of eutectic and the degree of segregation. As a result Scheil tends to underpredict the amount of solid above approximately 50% volume fraction of solid, this excess of predicted remnant liquid acts as a safety factor.

4.1 Microsegregation validation

To validate the diffusion model maps of microsegregation around a dendrite were acquired by EPMA. Typical maps for iron and niobium, taken from a welded piece of IN718, are shown in Fig 7.

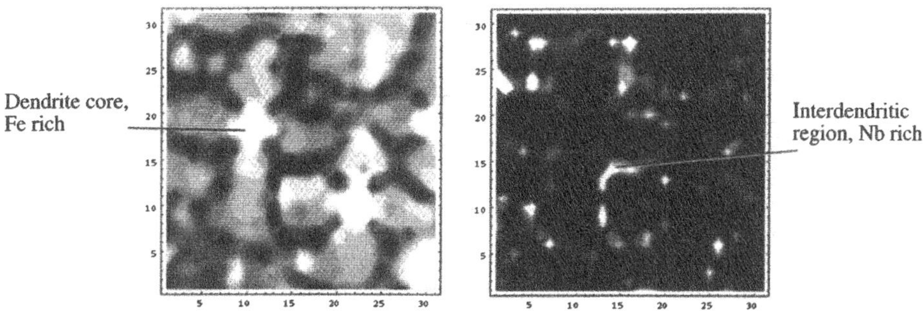

Fig 7. Composition profiles of Fe and Nb around a dendrite arm from IN718 TIG weld

As can be seen the iron remains predominantly in the dendrite core, whilst the niobium migrates to the interdendritic region. Knowing that the dendrite core represents the first solid formed and that the interdendritic represents the last remnant liquid we can use these two elements as a basis to measure the microsegregation of all other elements. These results are represented by the dots in Fig 8. The thin, solid lines in Fig 8. are the Scheil approximation to this segregation. As can be seen the absolute values deviate from those measured slightly, however the trends are correct and it should be remembered that Scheil represents a safe upper bound.

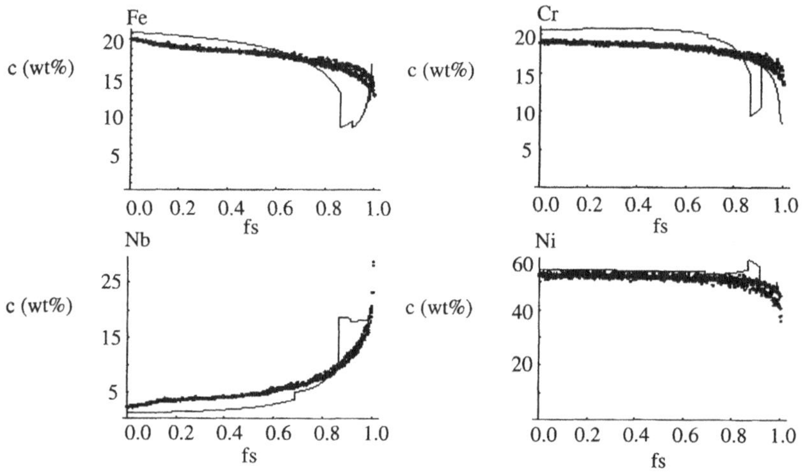

Fig 8. Comparison of measured profiles and Scheil (line) predicted

4.2. Application to solidification cracking

Having developed the Scheil methodology we calculate a 98% volume fraction solid at approximately 1000oC. If we now plot the history at a given point in the centreline of the weld path we can see if cracking will occur. Our thesis is that remnant liquid plus a tensile transverse stress will cause cracking to occur. Fig 8. shows the evolution of stress and temperature with time. For this example, the standard IN718 weld condition, the time at 1000oC, represented by the hatched region, does not coincide with any transverse (s22) tensile stress and hence no cracking is correctly predicted.

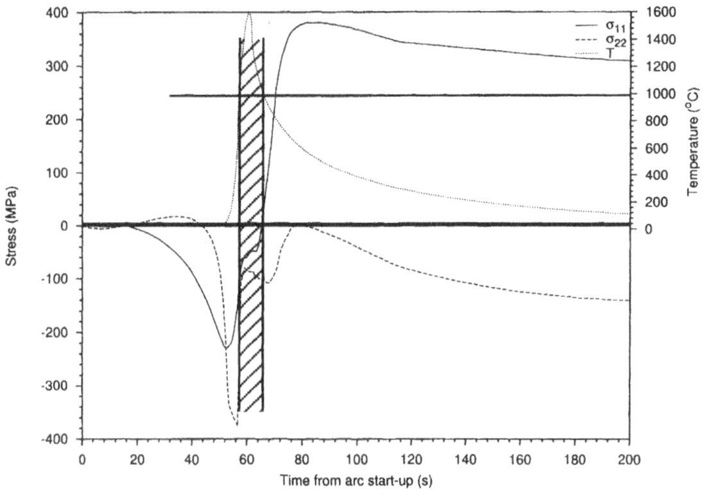

Fig 9. Transient stress (+ve tensile) and temperature distributions on the centre line of an IN718 TIG weld

5. CONCLUSIONS

Models have been developed for both TIG and EBW welding processes. The techniques have been applied to two nickel-base superalloys, IN718 and Waspaloy. Excellent agreement has been found between thermocouple measurements and finite element predictions for temperature evolution. Subsequent mechanical modelling has matched both measured distortion and residual stress. The Scheil approach has been coupled to such analyses to give an indication of the propensity of an alloy to crack under given welding conditions. Future work will be aimed at scaling up the models to the full size production components and investigating the influence of post weld heat treatment on the calculated residual stress field.

6. ACKNOWLEDGEMENTS

The authors acknowledge sponsorship from the Engineering & Physical Science Research Council (EPSRC), the Defence and Evaluation Research Agency (DERA), Farnborough and Rolls-Royce plc.

Mark Roberts and Roger Reed are employed in the Cambridge University Rolls-RoyceUTC, whilst Dave Dye is sponsored by EPSRC and Rolls-Royce, Olivier Hunziker is employed on an EPSRC/ DERA funded joint grant scheme - GR/L66380 and Howard Stone is employed on CEMWAMII (Cost Efficient Manufacture, Welding of Aerospace Materials) EPSRC grant - GR/M24127.

Informative talks with Paul Andrews, Bryan Benn, Douglas Crooke and others at Rolls-Royce, Mike Henderson at DERA, John Goldak at Carleton and Alan Oddy and Moyra McDill of OMNIS are also acknowledged.

For the residual stress measurements the use of the ISIS facility (RAL) Didcot UK, ILL facility Grenoble France, and Chalk River facility in Canada is appreciated. The advice of Colin Small, Phil Withers, John Bright, John Root, Tom Holden and Mark Daymond on diffraction techniques is also appreciated. Thanks also for the hole drilling performed by Phil Whitehead at Stresscraft and Gary Schajer at UBC for his advice in this area.

7. References

1. S.M. Roberts, H.J. Stone, J.M. Robinson, R.C. Reed: `Characterisation and Modelling of the Electron Beam Welding of Waspaloy', Mathematical Modelling of Weld Phenomena (4), H. Cerjak and H.K.D.H. Bhadeshia, IOM Communications, London ,1998, 631-648.

2. J.A. Goldak: `Coupling Heat Transfer, Microstructure Evolution and Thermal Stress Analysis in Weld Mechanics', Mechanical Effects of Welding, IUTAM Symposium, L.Karlsson et al, Springer-Verlag, 1992, 1-30.

3. L. Karlsson and L.E. Lindgren: `Combined Heat and Stress-Strain Calculations', Modeling of Casting, Welding and Advanced Solidification Processes V, M.Rappaz et al, The Minerals, Metals and Materials Society, 1991.

4. S.A.David, S.S. Babu, J.M. Vitek: `Weldability and Microstructure development in Nickel-base Superalloys' Mathematical Modelling of Weld Phenomena (4), H. Cerjak and H.K.D.H. Bhadeshia, IOM Communications, London ,1998, 269-289.

5. H.J. Stone, S.M. Roberts, T. Holden, P.J. Withers, R.C. Reed: `The Developmentand Validation of a Model for the Electron Beam Welding of Aero-engine Components', Trends in Welding Research, AWS,USA, 1998.

6. H.J. Stone, S.M. Roberts, R.C. Reed `Modelling of the Electron Beam Welding of Compressor Assemblies' Keynote speech 5th International Seminar on Numerical Analysis of Weldability, Graz, Austria. Proceedings to be published in Mathematical Modelling of Weld Phenomena (5), 1999.

7. D. Rosenthal: `The Theory of Moving Sources of Heat and Its Application to Metal Treatments', Trans. of the ASME, 1946, 849-866

8. A.S. Oddy, J.M.J. McDill, J. A. Goldak `Consistant Strain Fields in 3-D Finite Element Analysis of Welds'. Journal of Pressure Vessel Technology, 112, 309-311, 1990

9. O.C. Zienkiewicz, R.L. Taylor `The Finite Element Method: Solid and Fluid Mechanics, Dynamics and Non-linearity', volume 2, McGraw-Hill Book Company, Shoppenhangers Road, Maidenhead, Berkshire, England, 4th Edition, 1994.

10. D. Dye, S.M. Roberts, P.J. Withers, R.C. Reed: `The Determination of the Residual Strains and Stresses in a TIG-welded Sheet of IN718 Superalloy Using Neutron Diffraction', accepted for publication in the Journal of Strain Analysis, 1999.

11. H.J. Stone, P.J. Withers, T. Holden, S.M. Roberts, R.C. Reed: 'Characterisation of the Residual Stresses and Strains in an Electron Beam Welded Waspaloy Plate Using Neutron Diffraction, X-Ray Diffraction and Hole-Drilling', Metall & Mater Trans, 30a, 1797-1808 1999.

12. G.M. Gulliver: Metallic Alloys, Griffen, London 1922.

13. E. Scheil: Z.Metallkd, 1942, 34, p70.

14. O. Hunziker, D. Dye, S.M. Roberts, R.C. Reed `A Coupled Approach for the Prediction of Solidification Cracking During the Welding of Superalloys', 5th International Seminar on Numerical Analysis of Weldability, Graz, Austria Proceedings to be published in Mathematical Modelling of Weld Phenomena (5), 1999.

New capabilities of the software product SYSWELD for the prediction of welding residual stresses and distortions

G MANGIALENTI and **T CARTER**
SYSTUS International, Lyon, France

1 INTRODUCTION

SYSWELD is a software product dedicated to the prediction of welding residual stresses and distortions. Especially, it accounts for metallurgical transformations during welding and their influence on the mechanical behaviour of the welding joint. New capabilities have recently been added : the coupling of volume elements in the welding joint with shell elements in the rest of the structure, the automatic moving mesh refinement following the welding positionenables now to simulate longer welding joints, particularly of thin structures. In addition, the simulation of aluminium welding has been enhanced through new metallurgical models dedicated to aluminium alloys.

2 SYSWELD MAIN FEATURES

2.1 General architecture of SYSWELD

The physical phenomena involved in these processes, and their interactions, as integrated by SYSWELD, are summarized in the figure 1.

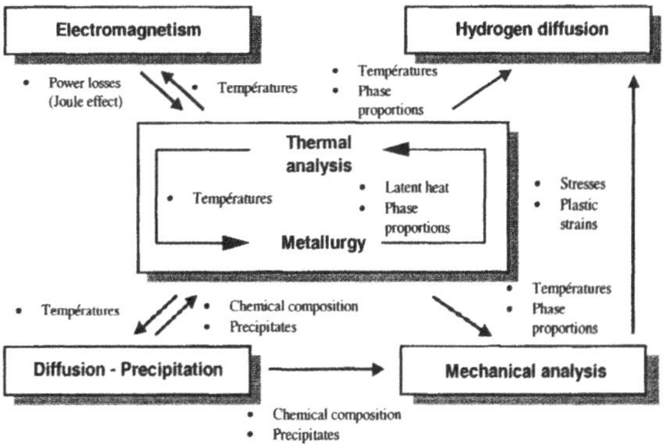

Figure 1 : *SYSWELD general architecture*

2.2 Basic capabilities

2.2.1 Thermometallurgy

There are three types of interaction between thermal and metallurgical analyses :

- metallurgical transformations depending directly on the thermal history of the part,
- metallurgical transformations accompanied by latent heat effects which modify temperature distribution,
- phase-dependent thermophysical properties.

A material is characterized at any moment, by the proportions of its constitutive phases. Different models [LeD], [PBRF] are available within SYSWELD to describe the transformation kinetics of most materials. The models can deal with several transformations between several phases.

For steels, the models are able to reproduce the Johnson-Mehl-Avrami law for transformations involving diffusion and Koistinen-Marburger law for martensitic transformation. The parameter of the model can be extracted from austenitization curve for transformation during heating and either a Continuous Cooling Transformation (CCT) diagram (figure 2) or an Isothermal Transformation (IT) diagram for transformations during cooling. The influence of austenitization conditions on transformations during cooling can also be accounted for via the austenite grain size computation.

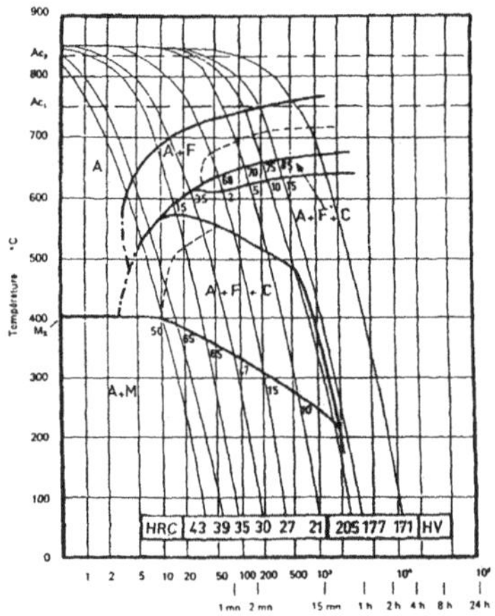

Figure 2. *Typical CCT diagram*

In the field of heat transfer, SYSWELD provides the user with very interesting features. Among other capabilities, the thermal properties can be both phase and temperature dependent, the latent heat effects due to transformations are accounted for using an enthalpy approach (figure 3) and a special module enables the modelling of radiation phenomena in enclosures.

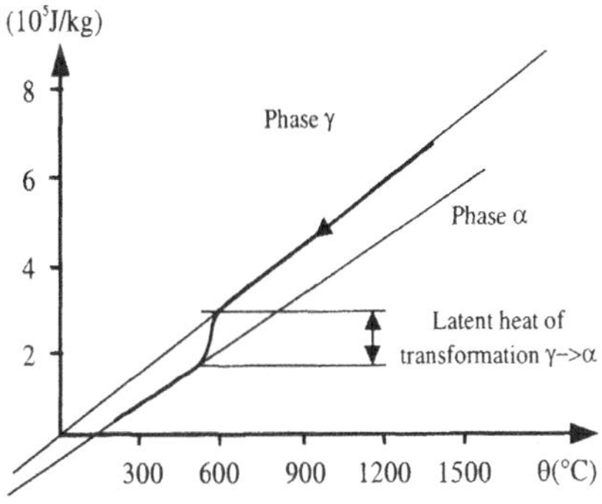

Figure 3. *enthalpy curves*

For welding processes, SYSWELD also provides a useful procedure to compute thermal, metallurgical and mechanical phenomena under steady state condition by solving the equations in the moving source frame [BPL].

2.2.2 Mechanical analysis

Metallurgy is involved in mechanical analysis, principally through two effects :

- volume changes resulting from metallurgical transformations,
- special behaviour linked to the multi-phase aspect of the material.

The volume changes occurring during metallurgical transformations are pointed out by dilatometry tests (figure 4). They are caused by modifications of the crystalline structure of the material during transformation, and are identified by free dilatometric tests. These changes are added to conventional thermal strain, and make a major contribution to the generation of residual stresses and strains.

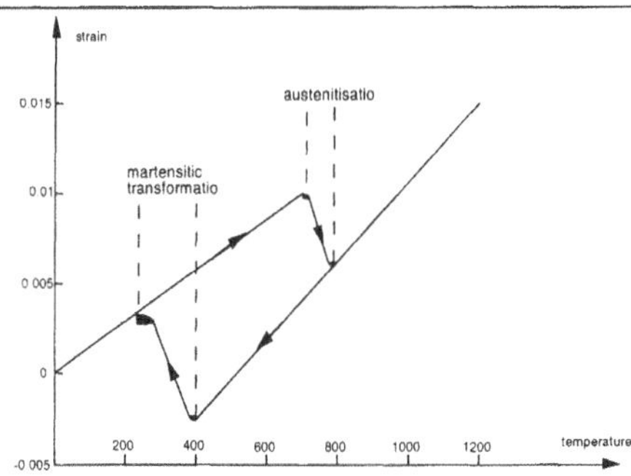

Figure 4. *Dilatometry test*

The behaviour law must be temperature-dependent, representative of the phase mix, and reproduce the so called transformation plasticity phenomenon (figure 5). SYSWELD includes a particularly sophisticated model for the elastoplastic behaviour of steels during transformations [LMD] [LDD]. The model can account for hardening and recovery (during transformations) effects.

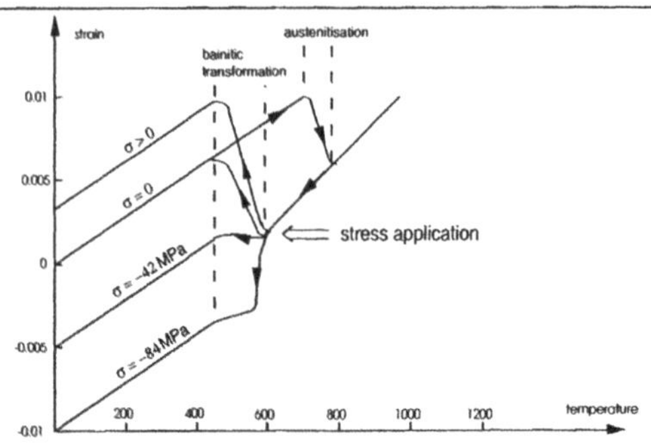

Figure 5. *Combined dilatometry test*

A wide choice of material behaviour law, including viscoplasticity, is also available in SYSWELD.

2.2.3 Hydrogen diffusion

Computation of hydrogen diffusion integrates the effects of temperature, metallurgical transformations and plastic stresses and strains. Modelling of these phenomena integrates trapping effects, and is of particular interest for the analysis of cold cracking risks.

2.3 Coupling of physical phenomena for resistance welding

Spot Welding requires the coupling of electrokinetic, thermal, metallurgical and mechanical phenomena. A fully coupled analysis of electrokinetic, thermal, metallurgical effects is available. For the coupling with mechanics, SYSWELD offers a tool to guide the step by step coupling, in order to update the contact areas during spot welding (electrode-sheet interface and sheet-sheet interface).

2.4 Welding dedicated tools

SYSWELD offers additional functions dedicated to continuous welding simulations.

The definition of any moving heat sources is available in SYSWELD. In order to further facilitate this task, three types of predefined heat sources are proposed. The first one is a Gaussian surface source dedicated to surface treatment, the second one is a conical gaussian volume source for the modelling of high energy welding processes (Electron Beam, Laser) and the last one is a double ellipsoidal source for conventional welding processes with material deposits. In all cases, the user only has to define few parameters such as the radius, the intensity or the velocity of the heat source.

The simulation of material deposition is automated using a refined elements birth procedure. This same procedure is available for the modelling of elements death in the simulation of cutting process.

3 NEW FEATURES

3.1 3D simulation of complex industrial component

This is a major market need to predict the residual distortions specially due to welding in order to optimize welding sequences.

Indeed, the prediction of residual stresses can be done on local models while the computation of global deformations requires the modelling of the whole structure. The mesh sizes involved with this kind of simulation are very important due the movement of the heat source and the very high temperature gradient near the weldment. This may lead to unreasonable CPU time.

SYSWELD offers different technics to avoid too large models. The first one is to make steady state simulations in the heat source comoving frame [BPL] as mentionned above. Two major additional capabilities have been recently added. They are presented below.

3.1.1 Shell-volume coupling

It is now possible to couple in the same analysis shell and volume elements. Metallurgical transformations are not taken into account in shells. However, this new feature enables to simulate more efficiently the welding of thin structures as the volume model can be limited to an area close to the heat affected zone, the rest being modeled with shell elements. Non linear thermal and mechanical behavior is available in shells. The computation is optimized so that no degree of freedom is added for volume elements. Compatibility elements are defined to keep and transfer rotations.

The following example shows the application of this capability to the laser welding of a thin steel sheet assembly.

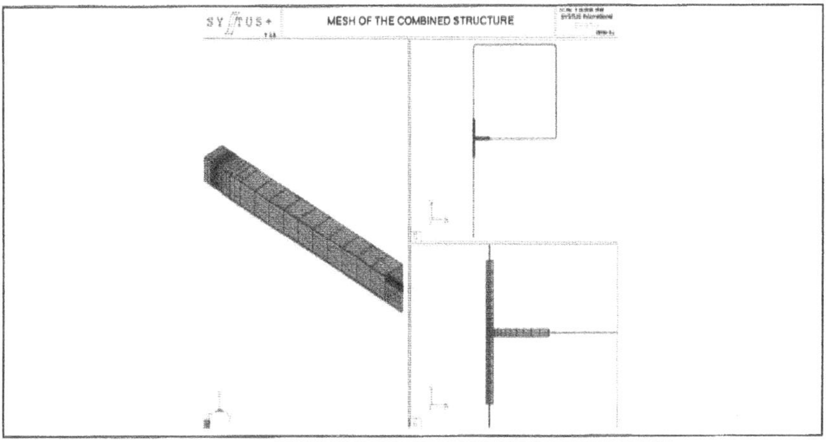

Figure 6. *Mesh*

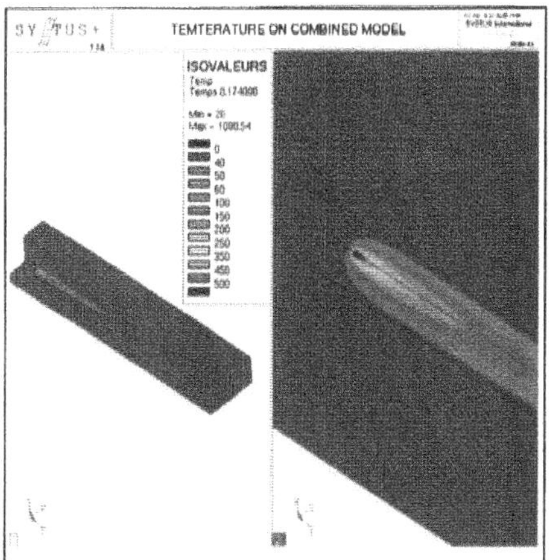

Figure 7. *Temperature field during laser welding*

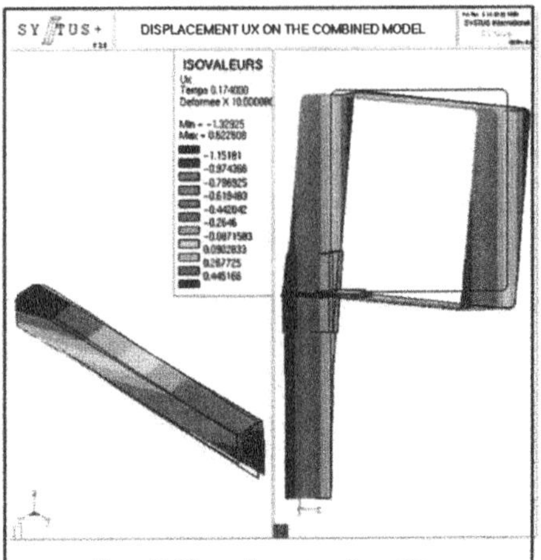

Figure 8. *Distorted geometry after welding*

3.1.2 Moving mesh refinement

An adaptive procedure based on a step by step analysis on a locally refined has been developed for applications such as welding, involving a moving heat source.

The user generates quickly a coarse mesh without taking into account the mesh refinement locally necessary for the considered analysis. He performs the whole pre-processing with this mesh (definition of material properties, boundary conditions, welding heat source...). Then, thanks to the moving mesh refinement function, the mesh can be regularly updated in order to use an optimized mesh, only refined around the current welding position.

At each mesh update, results are transfered from the old to the new mesh. The principle of the method is described in figure 9.

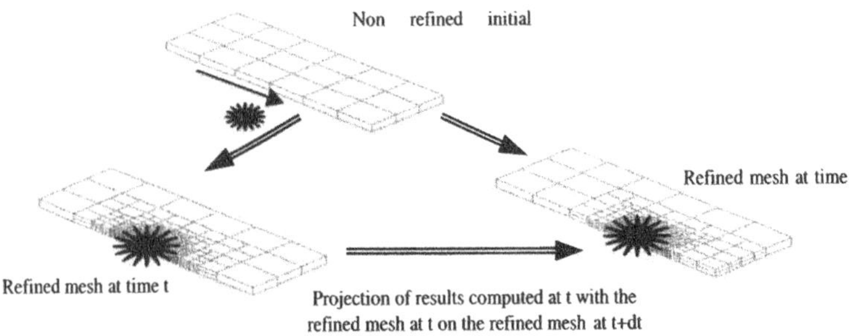

Figure 9. ***Principle of moving mesh refinement***

In the following example, moving mesh refinement is applied to laser surface treaptment of a steel plate. Due to symmetries of the structure, only a half-plate is considered in the simulation.

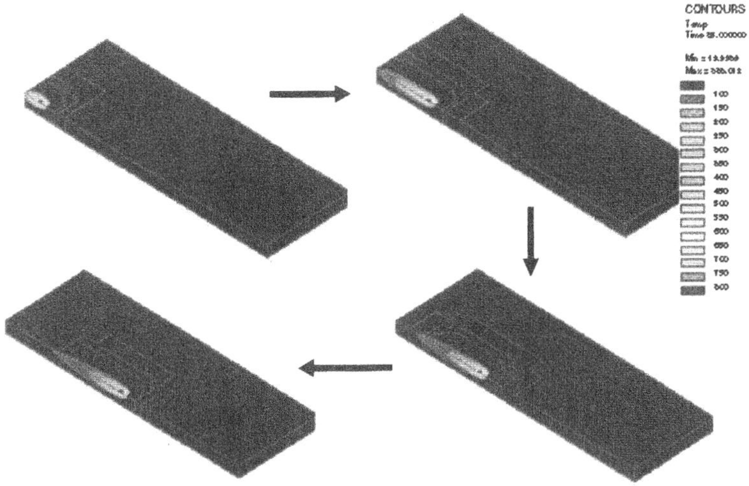

Figure 10. *Temperature field during laser surface treatment*

3.2 Modelling of aluminium alloys

Two metallurgical models have been developed :

- The first model is dedicated to alloys which can be hardened by heat treatment (precipitation): 2000, 6000 and 7000 series. This model allows the simulation of a precipitate dissolution kinetics.
- The second model is dedicated to alloys which can be hardened by strain hardening (cristallization) : 1000, 3000, 4000 and 5000.series. This model allows the simulation of a recrystallization kinetics.

During welding, the heating at high temperature locally generates recristallisation for work hardened alloys and dissolution of precipitates for alloys hardened by precipitation. From a mechanical point of view, the generation of the new structure in the heat affected zone is, in the two cases, accompanied by a strong decrease of properties.

To model these phenomena in SYSWELD, two fictitious phases are considered : an initial hard phase and a generated soft phase. The hard phase corresponds to precipitation or work hardened alloy. The soft phase corresponds to the recristallized or "de-precipitated" structure. During the thermo-metallurgical analysis, a transformation from the hard to the soft phase is defined, depending on temperature and time. During the mechanical analysis, the mechanical properties of the hard and soft phases are different.

The model of precipitate dissolution modell has been used for example for the simulation of the MIG welding process of a motorcycle rim made out of anticorodal, in co-operation with the Tüv Automotive GmbH.

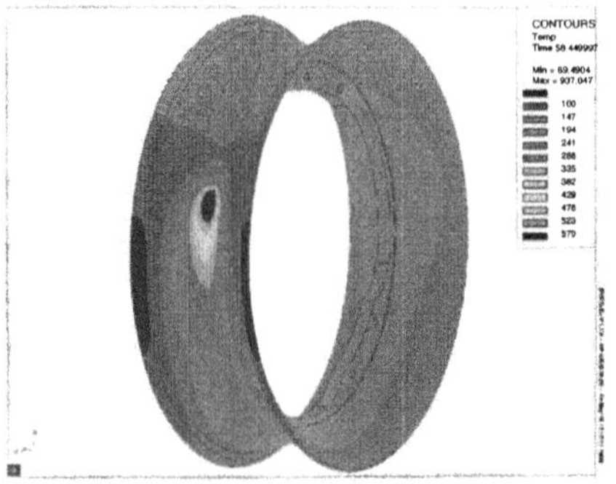

Figure 11. *Temperature field at the end of welding*

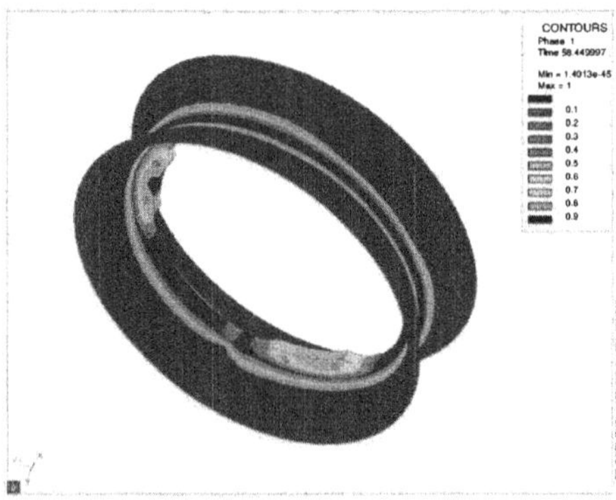

Figure 12. *proportion of precipitate dissolution during welding*

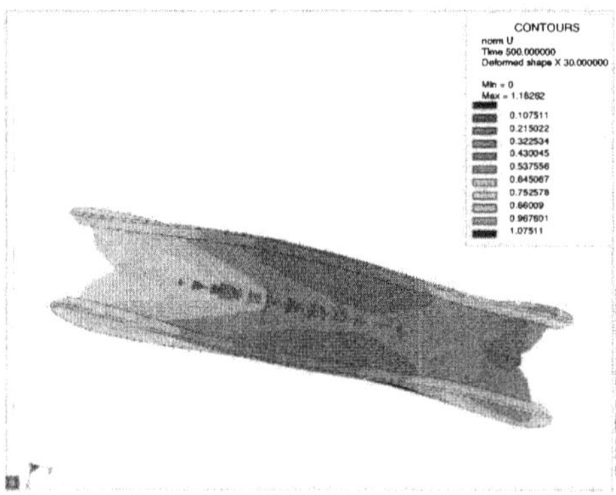

Figure 13. *Residual distortions after welding*

4 CONCLUSION

The SYSWELD technology has been validated by numerous welding applications, for several years. New capabilities exposed above, coupled with high performance computers enable now to perform simulations at an industrial stage. Real structures instead of simplified models can be studied in a reasonable time. Therefore, parts can be optimized through enhanced fatigue or fracture analysis that include welding residual stresses in addition to in service loadings.

However, the welding simulation of large thin structures is still difficult to handle through volume element modelling. This is why new simplified methods for the quick prediction of distortions during assembly processes are being investigated now, in order to optimize welding sequence with regard to the minimization of distortions.

Bibliography

[BeL] J.M. Bergheau and J.B. Leblond, « Coupling between heat flow, metallurgy and stress-strain computations in steels - The approach developed in the computer code SYSWELD for Welding or Quenching », Proc. of Vth Eng. Found. Conf. on Modeling of Casting, Welding and Advanced Solidification Processes, pp 203-210, Davos (Switzerland), Sept., 1990

[BPL] J.M. Bergheau, D. Pont and J.B. Leblond, « Three-dimensional simulation of a LASER surface treatment through steady state computation in the heat source comoving frame », Proc. of IUTAM Symp. on Mechanical Effects of Welding, Lulea (Sweden), pp 85-92, June 1991

[Hau] H. Häuser, « FE simulation of welding processes with 5XXX and 6XXX alluminium alloys by taking into account effetcs of recrystallization, precipitation reversion and natural ageing », 4th Int. Seminar on Numerical Analysis of Weldability, Seggau near Graz, 29. Sept. - 1. Oct. 1997

[LDD] J.B. Leblond, J. Devaux and J.C. Devaux, « Mathematical modeling of transformation plasticity in steels - I : Case of ideal-plastic phases - II : Coupling with strain hardening phenomena », Int. Jour. of Plasticity, Vol. 5, 1989

[LeD] J.B. Leblond and J.C. Devaux, « A new kinetic model for anisothermal metallurgical transformations in steels including effect of austenite grain size », Acta Metallurgica, 32, 1984

[LMD] J.B. Leblond, G. Mottet and J.C. Devaux, « A theoretical and numerical approach to the plastic behaviour of steels during phase transformations - I : Derivation of general relations - II : Study of classical plasticity for ideal plastic phases », Jour. Mech. and Phys. of Solids, Vol. 34, n°4, 1986

[PBRF] D. Pont, J.M. Bergheau, M. Rochette, R. Fortunier, « Identification of a kinetic model for anisothermal metallurgical transformations in steels », Proc. of 2nd Int. Symp. on Inverse Problems in Eng. Mech., pp 151-156, Paris, Nov., 1994

[Sar] E. Sarrazin, « Modélisation du soudage d'alliages d'aluminium », Thèse de Doctorat de l'Ecole Polytechnique., 1995

S708/005/2000

Measurement and prediction of residual stresses in a thick section steel welds

D J SMITH and **D GEORGE**
Department of Mechanical Engineering, University of Bristol, UK
P J BOUCHARD
British Energy, Barnwood, UK
C WATSON
Rolls Royce Marine, Derby, UK

ABSTRACT

The aim of this paper is to review recent developments in the measurement and prediction of residual stresses in thick section steel welds. Particular reference is made to the use of the deep hole method for measurement of residual stress. Results are presented from detailed measurements and predictions of the through thickness residual stresses in stainless steel welds. Some of the components are of complex shape and these thick section components are particularly challenging owing to the complexity of the geometry, both for the application of the deep hole method and finite element analysis. Residual stress predictions were made using an ABAQUS based finite element simulation of multi-pass welding. The similarities between the measurements and predictions provide increased confidence in the measurement technique and the use of FE as a predictive tool.

Note: Substantial parts of this document have either been published in a conference paper at ECRS-5 in Holland, September 1999,or are being considered for a Special Issue of the Journal of Strain Analysis on Residual Stresses. This paper also provides more recent results of our work.

1. INTRODUCTION

Residual stress measurements are most readily carried out on the surface of a component at start of life or after service duty. The techniques commonly used for these measurements include X-ray diffraction and various hole-drilling techniques [1-4]. For stress measurements through the thickness, the neutron diffraction technique [5] can be used. This has the capability of penetrating to a depth of about 25mm in steel components, providing access to both surfaces is available and the component can be taken to a reactor source. Other techniques, such as sectioning [6], destroy the specimen completely. Recent work at the University of Bristol has focused on using the Deep Hole (DH) method [7-9] for obtaining through thickness distributions of residual stress.

Published measurement methods for quantifying residual stresses in large section steel weldments are sparse. This is because measurement of the through thickness distribution of in-plane residual stresses in thick-section steel welds can be achieved by only few of the above techniques.

The deep hole method initially devised by Beaney [10] and Zhdanov et al [11], measures through thickness residual stresses semi-destructively. The method is classified as such because a small cylindrical core through the thickness is extracted, while the remainder of the component remains intact. Following on from earlier developments made by Procter et al [12], and Mitchell [13], significant improvements to the DH method have been achieved. Leggatt and co-workers, [14], using the DH method, carried out independent studies. A calibration study was undertaken using steel rectangular bar subjected to plastic deformation and unloaded. The method has been applied to obtain through thickness residual stresses distributions in butt-welded ferritic steel 85mm thick walled pipes [15,16] in the as-welded and post-weld heat treated conditions, and in butt-welded 100mm and 50mm thick steel plates [17]. Moreover, the deep hole method can be applied to more complex shaped components than just cylinders and plates

Experimental measurement of residual stress is not always a preferred option, although it should be utilised as validation for analytical and numerical studies. Nevertheless, finite element (FE) systems are used increasingly to simulate material response and the generation of residual stresses during welding [18]. One of the advantages of the FE approach is that it allows parametric studies to be carried out. For example, Brickstad and Josefson, [19] performed a series of FE analyses for butt welds in stainless steel pipes up to 40mm thick and compared the results with ASME Task Group recommendations [20]. Finite element analyses also provide three-dimensional residual stress states, which can aid understanding and characterisation of incidences of failure. For example, Holt and Spindler, [21] used the results from a welding simulation to simulate stress relaxation and creep damage under elevated temperature service conditions.

The aim of this paper is to review recent developments in the use of the deep hole measurement method. Measured stresses for a variety of steel components are presented to demonstrate the versatility of the method. Also presented are finite element predictions of residual stresses in thick section steel welds. Specifically detailed studies carried out to measure and predict through thickness residual stresses in AISI type 316H austenitic stainless steel butt-welded cylinders, cylinder-to-nozzle welds, and an end cap welded on a stainless steel cylinder. The thick section cylinder-to-nozzle and end cap welds are particularly challenging owing to the complexity of the geometry, both for the application of the DH method and for the FE model required. Residual stress predictions were made using an ABAQUS based finite element simulation of multi-pass welding, and subsequent service exposure. The predictions are compared with DH measurements at selected locations in and adjacent to welds.

2. DEEP HOLE METHOD

2.1 Basic Principles and Procedures

The deep hole method measures the distortion of a reference hole drilled through the component. A schematic of the operations is shown in Figure 1. Prior to drilling the reference hole, small steel blocks (or bushes) are glued to the surfaces of the component. The reference hole is drilled through the component and the bushes. Accurate measurement of the diameter of the reference hole is made at a number of angles. Then a column of material containing the reference hole as its axis is trepanned free of the component. Any changes in the axial dimension of the column are also measured. This can be done provided there is access for displacement transducers placed on the opposite side of the component wall, as shown in Figure 1. After trepanning the reference hole is re-measured, and the changes in diameter and column height are used to determine the residual stresses. Leggatt et al, [14] and Bonner and co-workers, [15-17] describe further details of the process.

The experimental equipment comprises six major components. A substantial frame supports large test specimens and provides vibration isolation. A common reference frame is used to locate four items of equipment, a gun-drill, a measurement system and an electro-discharge machine (EDM) for trepanning the column containing the reference hole. Finally, there is a computer system for controlling various aspects of the machining and for logging measured displacements and temperatures.

The gun-drill machines the reference hole. The measuring system obtains measurements of the diameter of the reference hole using a "Mercer" air probe. The height variation of the column during trepanning is obtained using capacitance gauges and thermocouples measure variations in temperature. In carrying out experiments, it is essential that careful alignment be made between the component and the reference frame. This is done using laser alignment. Earlier work by Bonner, [17] used an electro-chemical machining process for trepanning the core. This has been replaced by an EDM system to improve the co-axiality between the reference hole and the trepanned core, as shown in Figure 1.

2.2 Analysis

Trepanning the column containing the reference hole releases residual stresses. The experimental method measures the change in diameter of the reference hole and the change in length of the column arising from the trepanning operation. These measurements are converted to strains and used in the following analysis to determine the variation in residual stresses along the axis of the reference hole. Full details of the method of analysis, can be found in Bonner, [17], and Leggatt et al. [14]. More recent developments are also reported by Garcia Granada et al, [22]. Note that it is assumed that the axis of the reference hole is a principal stress direction and that the stress relaxation at the hole edge is elastic.

In the DH method, the radial displacements, u_r, are measured as a function of angle, θ, of the reference hole, (of diameter, d_o), and through-thickness position, z. The change in thickness is measured also when residual stress is released as a result of trepanning.

The relaxed normalised displacements, $2u_r/d_o$ are therefore related to the residual stresses, σ_{xx}, σ_{yy}, τ_{xy}, where

$$\left.\begin{aligned} \varepsilon_\theta[\theta,z].E = E\frac{2u_r}{d_0} = -\{\sigma_{xx}.f[\theta,z]+\sigma_{yy}.g[\theta,z]+\tau_{xy}.h[\theta,z]-\upsilon.\sigma_{zz}\} \\ \varepsilon_{zz}.E = -\sigma_{zz} \end{aligned}\right\} \quad (1)$$

The functions f [θ,z], g [θ,z], and h [θ,z] are given by Garcia-Granada et al [22]

At a given through thickness position, z_l, a minimum of four strains must be measured to obtain the residual stresses. For example $\varepsilon_\theta[\theta,z_l]$ at $\theta = 0°$, 45° and 90° and also ε_{zz}. Alternatively, as proposed by Bonner, [17], and Bonner and Smith [15], if measurements are obtained at n angles a least squares fit to the diametral strains can be used to determine the stresses. For a given through thickness position z_l equation (1) is rewritten in matrix form as :

$$\bar{\varepsilon} = -[M].\bar{\sigma} \quad (2)$$

where the strain and stress vectors are

$$\begin{aligned} \varepsilon &= [\ \varepsilon_\theta\ [\theta_1,z_l],\ \varepsilon_\theta[\theta_2,z_l],\ \ldots,\ \varepsilon_\theta[\theta_n,z_l],\ \varepsilon_{zz}\ ]^T \\ \sigma &= [\ \sigma_{xx},\ \sigma_{yy},\ \tau_{xy},\ \sigma_{zz}\ ]^T \end{aligned} \quad (3)$$

The compliance matrix M is given by

$$[M] = \begin{bmatrix} f[\theta_1,z_l] & g[\theta_1,z_l] & h[\theta_1,z_l] & -\upsilon \\ f[\theta_2,z_l] & g[\theta_2,z_l] & h[\theta_2,z_l] & -\upsilon \\ & & & . \\ . & . & . & . \\ f[\theta_n,z_l] & g[\theta_n,z_l] & h[\theta_n,z_l] & -\upsilon \\ -\upsilon & -\upsilon & 0 & 1 \end{bmatrix} \quad (5)$$

The optimum residual stress can be obtained by using Pseudo-Inverse or Moore-Penrose inverse matrices, where

$$\bar{\sigma} = -[M]^*.\bar{\varepsilon} \quad (6)$$

where $[M]^* = (M^T.M)^{-1}.M^T$ is the Pseudo-Inverse of the matrix [M] and $\bar{\sigma}$ is the optimum stress vector $[\sigma_{xx}, \sigma_{yy}, \tau_{xy}, \sigma_{zz}]^T$ that best fits the measured strains ε.

Residual stress measurements, using this analysis are described and illustrated in the next section.

3 MEASUREMENTS

3.1 Specimens

Measurements have been obtained from a variety of specimens and component shapes. These are shown schematically in Figure 2, and include stainless steel circumferentially butt-welded steel pipes, Fig 2a. The wall thickness of the pipes varied from 65mm to 19mm. These pipes were also used to study the influence of weld preparation and procedure, and the effects of post-weld heat treatment on the residual stress distributions. For pipes with 35mm and 19mm wall thickness, repair welds were introduced into the pipes and the through-thickness residual stress distributions measured.

Residual stress measurements have also been obtained in nozzle-to-cylinder welds, Fig 2b. The nozzle was part of a superheater header for a steam power generation system. The nozzle was welded onto the side of a cylinder, with most of the weld preparation made on the nozzle, and only a small amount of machining of the header. The header cylinder and outlet nozzles were both about 63.5mm thick. Prior-to-service (as-welded) and ex-service nozzles were examined. The ex-service nozzle had been exposed to an internal pressure of 16MPag, at 525°C for 55,000 hours.

For the ex-service nozzle, measurements were obtained through the thickness of the component at various locations. One important location was line A, which was through the cylinder. The line was orientated 60^0 to the axis of the nozzle coinciding with observed service degradation.

The end cap was welded to a stainless steel cylinder. Both were forged components. The cylinder was approximately 100mm thick with 500mm outside diameter. Submerged arc welding was used to join the end cap to the cylinder. Residual stress measurements were made in the centre of the weld in the as-welded condition.

3.2 Results

For all the specimens and components the procedure for measuring the through-thickness residual stress was essentially the same. In the following, examples are given of results obtained using the deep hole method, some of which have been presented in earlier papers. In addition, new measurements from the cylinder-to-nozzle and end cap-to-cylinder components are presented and compared with results from FE simulations.

Typical diametral strains obtained along the length of a reference hole are shown in Figure 3. The results were obtained from a butt-welded stainless steel 316H pipe of nominal thickness 35mm (but locally thicker). The axis of the reference hole was through the centre of the weld. Measurements of the diametral strain along the axis of the reference hole were obtained at 16 angular intervals of 11.25^0. For clarity, results are only shown for angles at 0, 45, 90 and 135^0 in Fig. 3a. The typical angular variation of the measured strain is shown in Fig. 3b for a depth of 15mm along the axis of the reference hole.

Stainless steel butt-welded cylinders

Experimental results obtained from butt-welded stainless steel cylinders of nominal wall thickness 64 and 35mm, outer radius 432mm and the same material are shown in Figs. 4 to 7. In the 64mm thick cylinders MMA and narrow gap welds were introduced. Results for MMA weld are shown in Figure 4. Results for the TIG welds compared with the MMA weld results in Fig. 5. Overall the measurements reveal that higher residual stresses occurred in the narrow gap TIG weld.

Distributions of residual stress in the hoop and axial directions are shown in Figure 6 for the original girth weld in the 35mm thick cylinders. The residual stresses were measured adjacent to the heat-affected zone of the weld. The largest measured stress occurred in the hoop direction at a position 7mm below the outer surface. At the inner wall the hoop stress was compressive. The peak axial residual stress was about half the value of the hoop residual stress.

In all these cylinders tensile residual stress towards the outer surface and compression near the inner surface. The tensile hoop residual stress was however larger in the 64mm thick cylinder. More notable are the differences in the residual stresses on the inner surfaces of the 64 and 35mm thick cylinders. The axial residual stress at the inner surface of the 35mm thick cylinder was in tension. For the 64mm thick pipe the hoop and axial residual stresses were close to zero on the inner surface.

The 35mm cylinder weld was then machined using a hand grinder to create three excavations: two to a depth of 25% and one to a depth of 75% the section thickness. The effective circumferential lengths of the weld excavations each corresponded to a subtended angle of approximately 20°. The excavations were filled with type 316L weld metal using a MMA process, creating two shallow 25% thickness repairs and one deep 75% thickness repair. Figure 7 shows the residual stresses measured, again in the HAZ of the original weld. The peak residual stress in the hoop direction was similar to the original weld but the repair weld has elevated the residual stress towards the inner surface of the cylinder. The axial residual stresses after introducing a repair weld was also increased resulting in a predominantly tensile distribution through the wall thickness of the cylinder.

Stainless steel welded cylinder-to-nozzle

For the nozzle-to-cylinder component, measurements were obtained in both the as-welded and post-service conditions. Results at location A for these two conditions are shown in Fig. 8. The orientation of the measured residual stresses is also shown in the figure. The largest measured residual stress occurred in the nozzle hoop-cylinder axial plane, normal to line A, towards the outer surface of the cylinder. This residual stress was tensile for depths up to about 25mm. Thereafter, it was compressive through to the inner surface of the cylinder. A similar pattern occurred for the residual stress, σ_y, along the axis of the cylinder, although the magnitude was lower than for the σ_x stress. For the post service condition, both σ_X and σ_Y stresses at the outer surface reduced to about one half their as-welded values, with the peak σ_x residual stress at about 200MPa near to the outer surface.

Stainless steel end cap-to-cylinder

Experimental results for measurements through the centre-line of the weld are shown in Figure 9. The highest residual stress occurred in the hoop direction of the cylinder and close to the outer surface of the cylinder. The peak residual stress was about 300 MPa. The peak axial stress was slightly lower at about 270MPa, but was tensile for about 25mm through the wall thickness.

4. FINITE ELEMENT ANALYSIS

4.1 Model Descriptions

All the finite element analyses were carried out using the ABAQUS code. For the 65 and 35mm thick cylinders axisymmetric models were used to determine the distribution of residual stress for the as-welded state. For the cylinder-to-nozzle component, full three-dimensional modeling of the welding process for this structure was judged to be impracticable. Instead, an axisymmetric idealisation of the component was created. The header cylinder was idealised as a sphere with the same internal diameter, and the set-on nozzle represented as an attached cylinder. The mesh for the whole model is illustrated in Fig. 10a, with a detailed section through the weld shown in Fig 10b. The elements were mostly hybrid 8 noded axi-symmetric reduced integration quadrilaterals, with a smaller number of 6 noded axi-symmetric triangles employed where necessary for mesh refinement.

A pragmatic approach to modeling the multi-pass weld deposits was adopted in order to reduce computational requirements. The entire weld was represented by 11 layers of filler metal, sequentially added to the model, one half-layer at a time. The layer thicknesses were based on an observed deposition structure from the cross section of a typical ex-service weld. However, each half-layer generally represented more than one weld pass. The half-layer 'lumped' beads were each divided into two elements in the thickness direction, and the element aspect ratios were kept as close to one as practicable. The resulting weld metal mesh is shown in shown in Fig 10b. The final 'lumped' beads represented the weld cap region. This material was partially removed, after the welding simulation, to model final grinding of the weld outer profile.

A similar approach was also used for the simulation of welding of the end cap-to-cylinder component. For the weld all 78 weld beads were introduced into the model. The weld bead size and numbers were estimated from a macrograph of a similar weld. The details of the FE mesh for the end cap-to-cylinder weld are shown in Fig.11.

For introducing the repair weld into the 35mm thick cylinder welds two analyses were carried out, a 3-D analysis and a detailed axisymmetric analysis. A three dimensional finite element analysis following the procedure described by Zhang et al [23]. A 3-dimensional (3-D) model of the pipe was constructed using composite shell elements and the as-welded residual stress-state from the axisymmetric analysis mapped around the entire circumference. Then, at defined locations, excavation of material was simulated using a layer activation/deactivation scheme. The introduction of repair weld metal was idealised using four "lumped" beads, rather than representing the many passes (>30) actually used.

An axisymmetric analysis, using the original axisymmetric model, was performed to simulate the detailed deposition of each repair weld bead. The stiffness of the axisymmetric model was modified to correctly represent the restraining effect of the unrepaired section. This was achieved by including a dummy beam tied to the model using multi-point constraints. The beam properties and constraints were derived from a 3-D shell model. The repair weld was simulated by first excavating original weld and parent material, and then adding new weld filler, bead by bead.

4.2 Material Properties

In all FE analyses temperature-dependent thermo-physical properties (thermal conductivity, specific heat, and density) for the parent and weld metal up to melting point (1420°C) were used in the analysis. These data were compiled from proprietary sources. Above this temperature, properties were held constant, except for the thermal conductivity. This was artificially increased to compensate for the effect of heat transfer due to convection stirring in the molten weld pool. Latent heat of fusion was ignored in order to improve numerical stability of the thermal analysis and because sensitivity studies using an alternative finite element code showed that it had a small effect on predicted temperatures.

Temperature-dependent thermal expansion, Young's Modulus, Poisson's Ratio and yield behaviour properties up to melting point were compiled for the mechanical analysis from proprietary sources, and tensile tests on ex-service header and weld materials. A rate independent, isotropic elastic-plastic material model with strain hardening up to 10% strain was used for the parent material. No attempt was made to model the complex visco-plastic behaviour of weld metal near to and above its melting point. For the service creep analysis, deformation was based on the RCC-MR creep law [24].

4.3 Residual Stress Analysis

A common approach was adopted for all the weld simulations, including simulation of the weld repairs. A thermal transient analysis was first carried out for the sequential addition of each lumped weld bead. The heat input was applied by assigning each element in the weld bead an initial temperature, and applying body and surface fluxes. A proprietary technique was used to calculate the appropriate total heat input for the lumped manual metal arc weld passes, and checked by comparing the predicted and actual depth of weld fusion side-wall penetration. Subsequent cooling, as the heat conducts into the parent material and is lost through surface convection and radiation. Convective and radiative heat losses from the surfaces of the nozzle and header were allowed for using standard heat transfer coefficient data. The cool down period after each weld deposition was limited to give a realistic interpass temperature. After the final bead had been analysed the model was reset to 20°C.

The mechanical analysis similarly modeled the sequential build-up of lumped weld beads or the detailed build-up of the weld beads depending on the component examined. All new bead elements were added to the model in an initial strain-free state. Loading was supplied from the temperature-time history of the thermal analysis.

measured. This is possibly due to the very high constraint introduced by the end-cap in the FE analysis, and may be much higher than is practically the case.

5. DISCUSSION

5.1 Measurement Method

The comprehensive range of components and wall thickness presented here reveal that the deep hole method for measuring residual stresses has wide applicability. There are nevertheless a number of assumptions and approximations, which have been made in using the analysis.

Using "stress free" steel blocks the accuracy of the method has been assessed as ±25MPa, (19). Further, it is assumed that the axis of the reference hole is one of the principal stress directions. If not, errors occur in transforming the measured diameter changes to stresses.

In some components, it has been possible to measure the change in length of the core as it is trepanned. However, the presence of thermal strains, during trepanning, make accurate determination of the residual stress relaxation difficult. Further work is being done to determine accurately the mechanical strain release.

5.2 Comparisons between Measured and Predicted Residual Stresses

In all the components there are a number of reasons for the differences between the predicted and measured residual stresses. Approximations and errors in the measurement method described above will give rise to uncertainty in the magnitude of the measured residual stresses. In addition, the finite element simulation relied upon a number of important assumptions in the thermal and mechanical analyses. In general a 2-D axi-symmetric idealisation of a complex 3-D geometry was assumed. For the simple cylinder butt-welds Figure 6 reveals that there is excellent agreement between the measurements and predictions.

In the case of the cylinder-to-nozzle component the axi-symmetric model's stiffness (and therefore its constraint on the weld) will certainly differ from that of a cylinder. Its resistance to the weld hoop stress distribution would be expected to be greater, leading to an over-prediction of stress component, σ_x in Fig 8a. Except for a region close to the outer surface of the component, this predicted stress is greater than measured. However, resistance to nozzle-axial loading in the axisymmetric model is likely to be similar to that of a cylinder (and a flat plate), leading to better predictions of σ_y stress.

Finite element studies have been performed to examine the influence of heat input, heat transfer properties and elastic-plastic properties on predicted stresses in thick section welds. The distribution of stress was found to be relatively insensitive to changes in thermal parameters. However, stress-strain behaviour at temperatures up to about 700°C was found to be more important. Other modeling assumptions, involving the lumping of beads and the use of axi-symmetry to model a three-dimensional welding process, are more difficult to assess. However, they might be regarded as

After the final weld run and removal of weld cap material, the residual stress-state corresponded to the as-welded condition at 20°C.

For the cylinder-to-nozzle weld thermal relaxation of the predicted residual stress field was investigated by performing a creep analysis under plant steady state operating conditions. The temperature of the as-welded model was first raised to 525°C and then an internal pressure of 16 MPag applied. A creep analysis was performed to simulate stress relaxation up to a time of 55,000 hours, followed by removal of operating pressure and reduction of temperature to 20°C.

4.4 Predicted Residual Stress Distributions

Figure 4 shows the predicted distributions for the as-welded residual stresses in the 64mm thick cylinder welds. Overall, the predictions gave larger residual stress distributions compared to the measurements, particularly in the hoop direction. Figure 6 also shows similar predicted distributions of residual stress for the original weld in the 35mm thick cylinder from the axisymmetric FE analysis. Overall, the agreement between the prediction and the experimental measurements is remarkably good bearing in mind the assumptions adopted in both the measurement method and the FE analysis. The shapes of the profiles match extremely well in each direction. The axial residual stresses differ near the inner surface, but this is not surprising because the weld root area was simplified in the FE analysis, the main interest being on correct modelling of the outer weld region.

Figure 7 compares experimentally measured stresses from the 75% repair, with predicted profiles from the 3-D and axisymmetric FE analyses for a deep weld repair with a 20° arc. There is good agreement between the shape of the measured distributions and simulated profiles from the 3-D analysis, although the latter predicts higher magnitudes of stress. In contrast, the axisymmetric, bead-by-bead, FE analysis for the deep repair predicts more detailed profiles with higher stress gradients. These are broadly consistent with the 3-D and experimental results, but do not closely match in detail. Both analyses predict a tensile net section axial stress at mid-length of the repair, as noted in the measured profile. The full results from the 3-D FE analysis showed how such transverse stresses local to the repair are equilibrated by compression beyond the ends of the repair.

Predicted stresses for the cylinder-to-nozzle component along the line at location A (see Fig. 10) are shown in Figure 8, for both the pre-service (as-welded) and post-service (thermally aged) conditions. Line A represents the centre-line of the reference hole for deep-hole residual stress measurements. The experimental results for both the pre-service and post-service conditions are also plotted on Figure 8. It is seen that the predicted residual stresses are generally greater than measured values. Towards the outer surface of the component, there is reasonably good agreement between measured and predicted residual stresses for both as-welded and post-service conditions. However, for regions where measured and predicted residual stresses were compressive, the predictions appear to overestimate the compressive stresses.

Predicted residual stresses for the end cap-to-cylinder weld are shown in Fig. 9. The agreement between measured and predicted axial residual stress distributions is excellent. For the hoop direction the predicted residual stresses are much higher than reasonable for predicting the overall structural behaviour of a thick section nozzle-to-cylinder weld, owing to the large number of beads actually used and the random stop/start positions of the manual metal arc welding process.

For the FE simulations of the weld repairs a number of simplifying assumptions cause the differences in results observed in Fig. 7 for the 75% weld repair. The detailed axisymmetric FE analysis indicates the presence of steep residual stress gradients through the wall of the pipe. In contrast, the 3-D composite shell analysis predicts a more uniform distribution of residual stress. This latter result is the consequence of using just 4 "lumped" beads. Furthermore a composite shell analysis only simulates the global response of the structure to the presence of the local repair. The good agreement between the experimental results and the 3-D analysis suggests that the measurement method using a 20mm core was also unable to detect steep residual stress gradients through the thickness. Further work is being carried out to assess the use of smaller cores to measure residual stress profiles with steep gradients.

6. CONCLUSIONS

a) The deep hole technique has been applied to a variety of thick section welded components, including pipes, nozzle-to-cylinder and end cap-to-cylinder components.

b) The through-thickness distribution of residual stresses has been obtained in components of wall thickness in the range 35mm to 108mm.

c) Finite element simulation of welding of a complex thick section structure has shown that residual stress distributions can be obtained using a combined thermal and mechanical analysis.

d) Measurements of through-thickness residual stresses using the deep hole method have revealed distributions remarkably similar to the predicted residual stress distributions.

e) The similarities between the predictions and measurements provide increased confidence in the use of finite element approach as a predictive tool and the use of the deep hole measurement technique as a means of verifying the predictions.

Acknowledgements

The financial support for this work was provided by a number of grants, including EPSRC GR/G 25399, funding from British Energy and the Industrial Management Committee (HSE(NII), British Energy and BNFL) and funding from Rolls Royce Marine. Their support and encouragement is gratefully acknowledged. Dr Martyn Pavier (at Bristol University), Dr Rick Leggatt (at TWI), Dr Neil Bonner (now at Rolls Royce) and Dr Peter Holt (previously at British Energy) have provided valuable advice throughout this work. We also grateful to Drs Zhang and Dong from Battelle for allowing us to use their 3D results in this paper. This paper is also published by kind permission of British Energy plc, and Rolls Royce plc.

References

1 K. Nagai and M. Iwata, 1984: "*Non Destructive Measurement of Residual Stress in a Filled Welded Joint by Photographic X-Ray Diffraction Technique*", IIW/IIS, Doc X – 1065 - 84.

2 Y. Ueda and K. Fukuda, 1989: "*New Measuring Method of Three Dimensional Residual Stress in Long Welded Joints Using Inherent Strains as Parameters – L_z Method*", Journal of Engineering Materials and Technology, vol. 111.

3 F. Faure and R.H. Leggatt, 1996: "*Residual Stresses in Austenitic Stainless Steel Primary Coolant Pipes and Welds of Pressurised Water Reactors*", Int. J. PVP, no 65, pp 265-275.

4 H.J. Yen, M.C.C. Lin and L.J. Chen, 1994: "*Measurement and Simulation of Residual Stress in Type 304 Weld Overlay Stainless Steel Pipe*", Int. J. PVP, no 60, pp 271-283.

5 T.M. Holden, J.H. Root, R.A. Holt and G. Roy, 1988: *Proc. 7th Int. Conf. On Offshore Mech. and Arctic Eng.*, pp 127-131.

6 Y. Ueda, K. Fukuda and S. Endo, 1975: *Trans. Jap. Weld. Res. Inst.*, vol. 4, no. 2, pp 13-27.

7 E. Proctor and E.M. Beaney, 1987: "*Advances in Surface Treatments: Technology-Application-Effects*", vol. 14, International Guidebook on Residual Stresses: 165-198, Oxford, Pergamon.

8 R.H. Leggatt, D.J. Smith, S.D. Smith and F. Faure, 1996: "*Development and Experimental Validation of the Deep Hole Method for Residual Stress Measurements*", J. Strain Analysis, vol. 31, pp 177-186.

9 N.W. Bonner and D.J. Smith, 1996: "*Measurement of Residual Stresses Using the Deep Hole Method*", PVP, vol. 327, pp 53-65.

10 Beaney, E.M. 1978. "Measurement of sub-surface stress". CEGB Report No. RD/B/N4325.

11 Zhdanov, I.M. & A.K. Gonchar 1978. Automatic Welding Vol.31 No.9: 22-24.

12 Procter, E. & E.M. Beaney 1987. Advances in Surface Treatments: Technology-Application-Effects, Vol.4. International Guidebook on Residual Stresses: 165-198. Oxford: Pergamon.

13 Mitchell, D.H. 1988. "R6 validation exercise: through thickness residual stress measurements on an experimental test vessel rig". CEGB Report No. RD/B/6088/R88 (Unrestricted).

14 Leggatt, R.H., D.J. Smith, S.D. Smith and F. Faure, 1996, "Development and experimental validation of the deep hole method for residual stress measurement", Journal of Strain Analysis, Vol. 31, pp 177-186.

15 Smith, D.J. and N. W. Bonner, 1994, "Measurement of residual stresses in a thick section steel weld", in Engineering Integrity Assessment, Edited by J.H. Edwards, J. Kerr and P. Stanley, (EMAS, Chameleon Press), 259-274

16 Bonner, N.W. and D.J. Smith, 1996, " Measurement of residual stresses using the deep hole method", PVP vol. 327, pp53-65.

17 Bonner, N.W., 1996, "Measurement of residual stresses in thick section steel welds", PhD. Thesis, University of Bristol.

18 Smith, S.D., 1991 " A review of numerical modelling of fusion welding for the prediction of residual stresses and distortion", TWI Report 437, December

19 Brickstad, B. and L. Josefson, 1996, "A parametric study of residual stresses in multi-pass butt-welded stainless steel pipes", Int. J. Press. Ves. & Piping, (75), 1998, pp 11-25.
20 ASME, 1986, Code Section XI Task Group for Piping Flaw Evaluation " Evaluation of Flaws in Austenitic Steel Piping", J. Pres. Ves. Technology, Vol. 108, 352-366
21 Holt, P.J and M.W. Spindler, 1996, " A practical application of continuum damage mechanics to plant integrity assessment", Symp. on Elasticity and Damage in Solids Subject to Microstructural Change, Newfoundland, September.
22 A.A. Granada-Garcia, D. George and D.J. Smith, "Assessment of distortions in the deep hole technique for measuring residual stresses", in Proceedings of the 11th. Exp Mech Conf, Oxford, Aug, 1998, pp1301-1306
23 J. Zhang, P. Dong and J.K. Hong, 1997: "*A 3D Composite Shell Element Model for Residual Stress Analysis in Multi-Pass Welds*", Trans. 14th Int. Conf. on Structural Mechanics in Reactor Technology (SmiRT-97), Lyon, France, vol. 1, pp. 335-344.
24 RCC-MR, 1985, Section 1, sub-section Z, Technical Appendix, AFCEN, Paris, 1985.

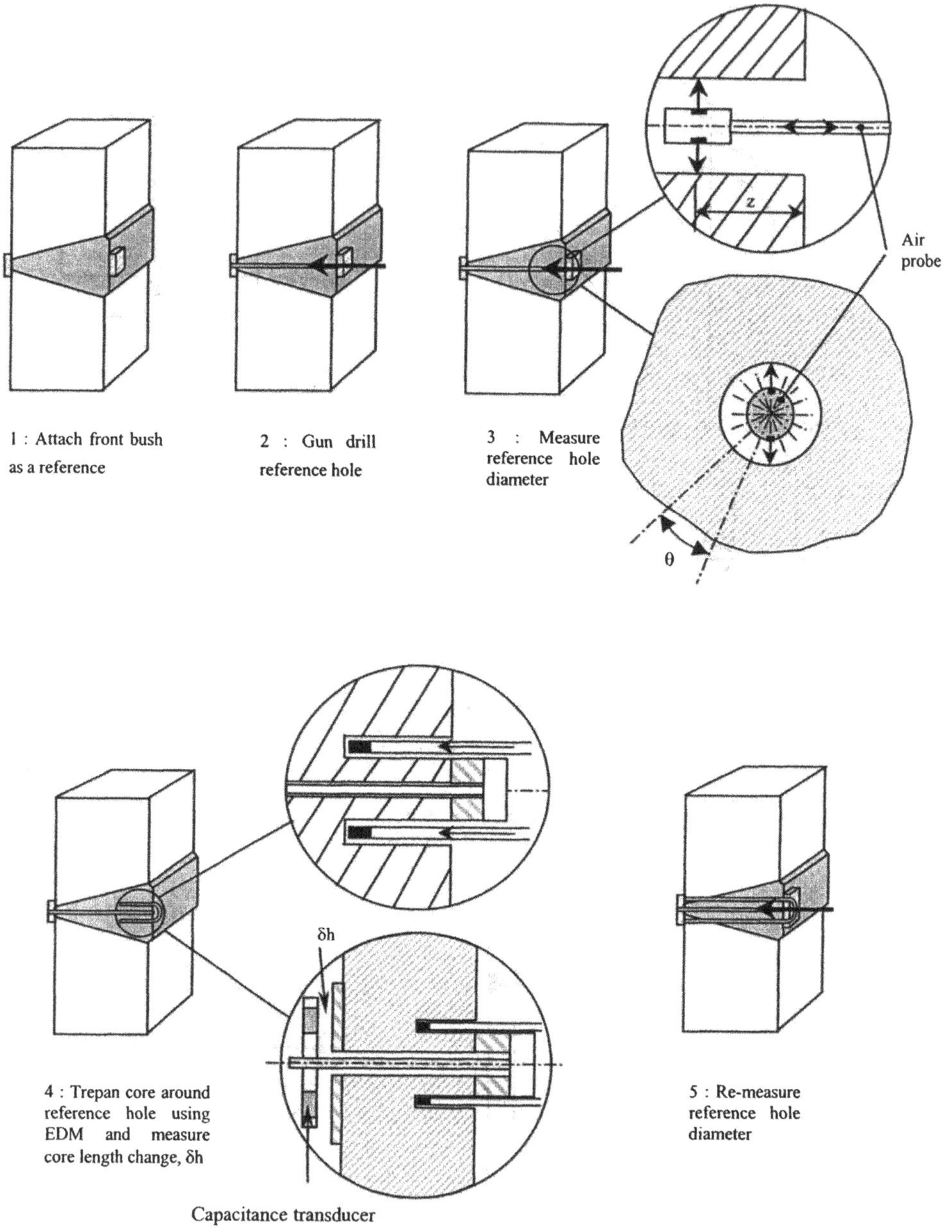

Figure 1 : Basic experimental process used in the deep-hole measurement method.

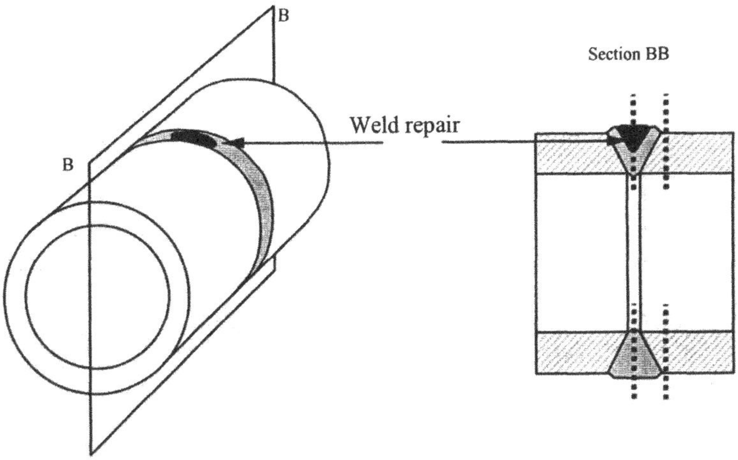

Figure 2a : Circumferential butt-welded pipes

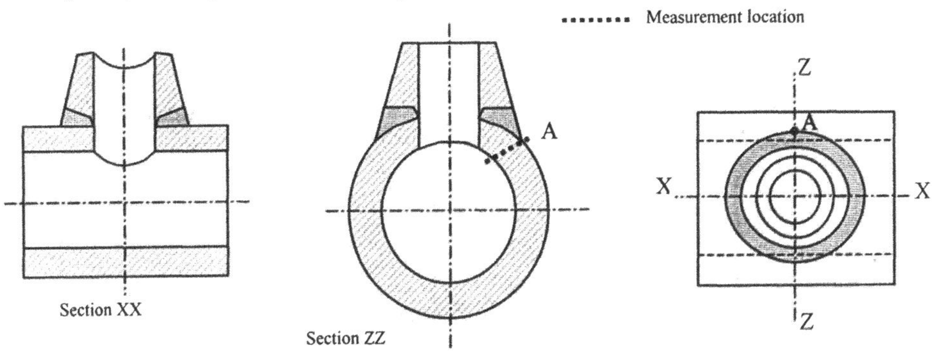

Figure 2b : Nozzle-to-cylinder welded component

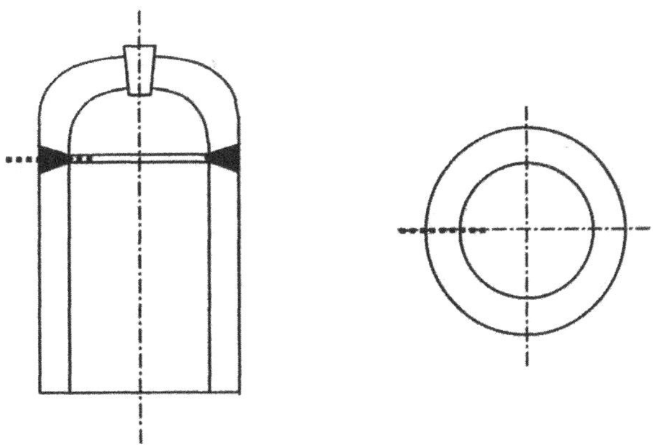

Figure 2c : End cap-to-cylinder component

Figure 2 : Components subjected to the Deep Hole Measurement Method

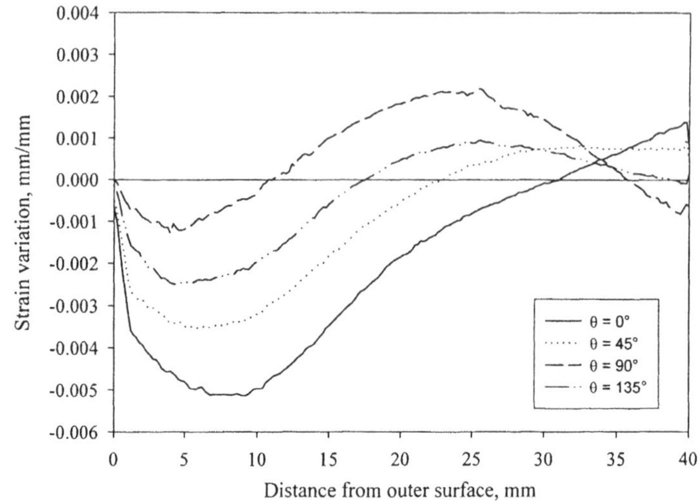

a) Variation of through thickness strains

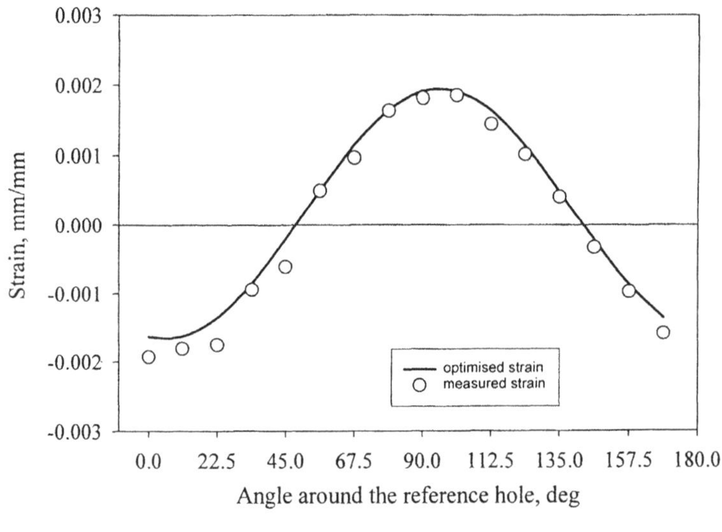

b) Measured and optimised strain at depth of 15 mm

Figure 3 : Examples of the measured strain variation resulting from deep hole measurement in a 35 mm thick butt-welded stainless steel pipe

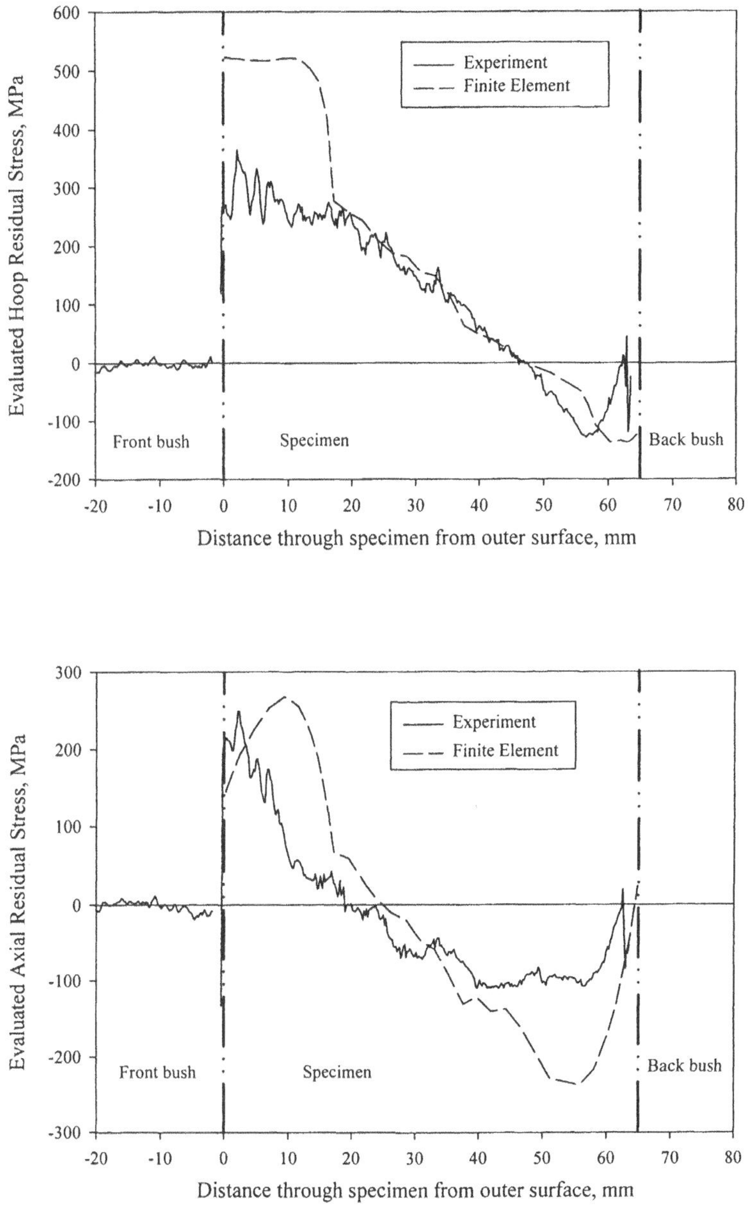

Figure 4 : Comparison of experimental and FE results through weld centre line of butt-welded MMA specimen.

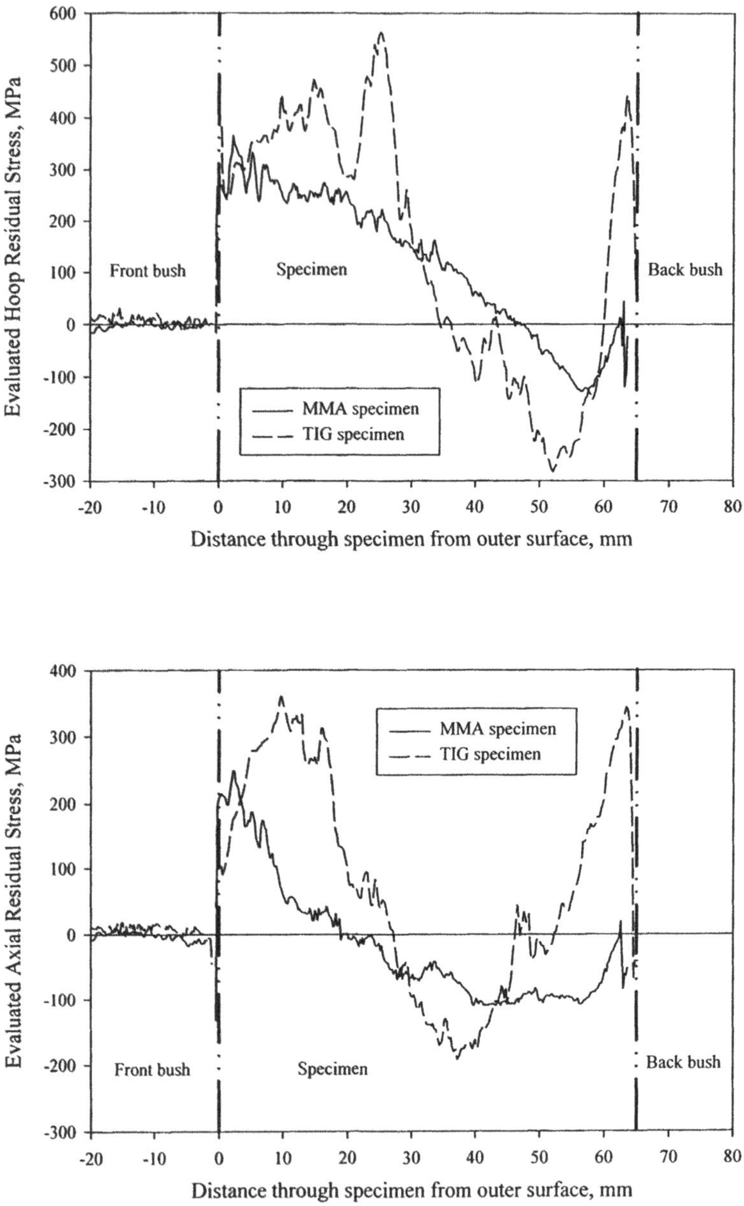

Figure 5 : Comparison of experimental results through weld centre line of butt-welded MMA specimen and narrow gap TIG.

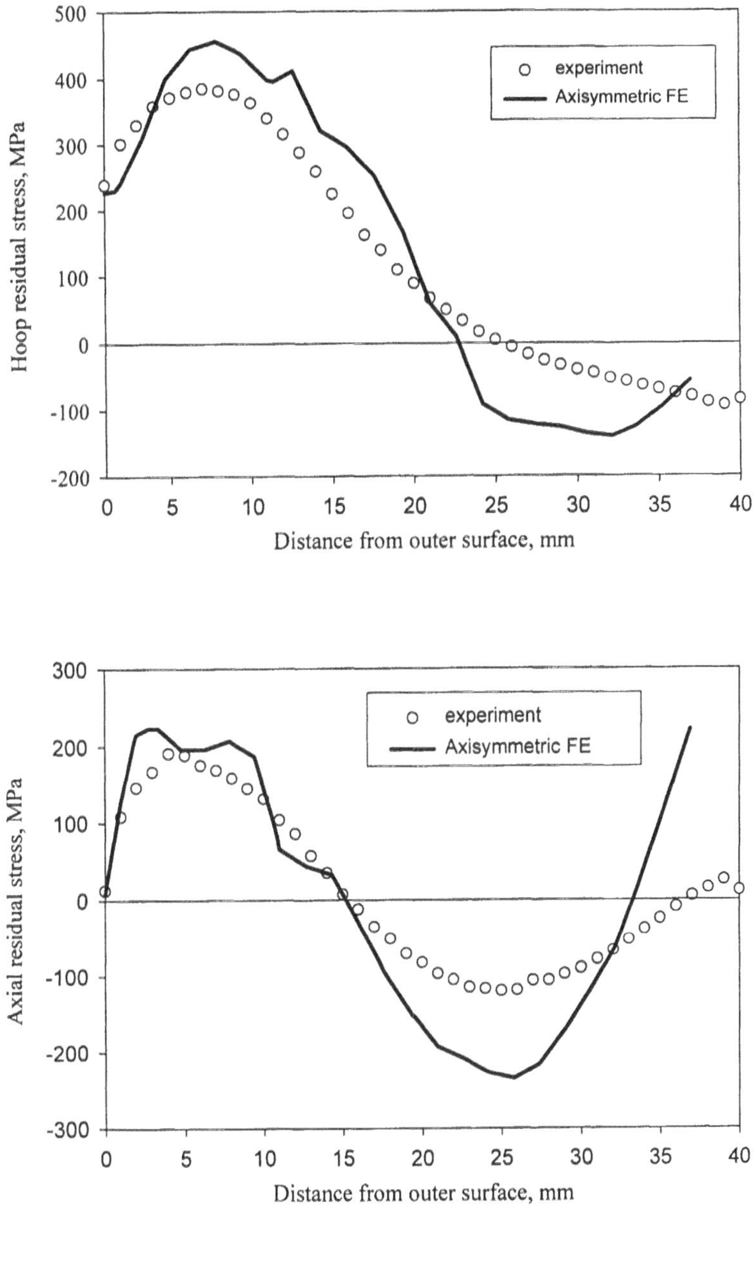

Figure 6 : Residual stress distribution in HAZ without weld repair

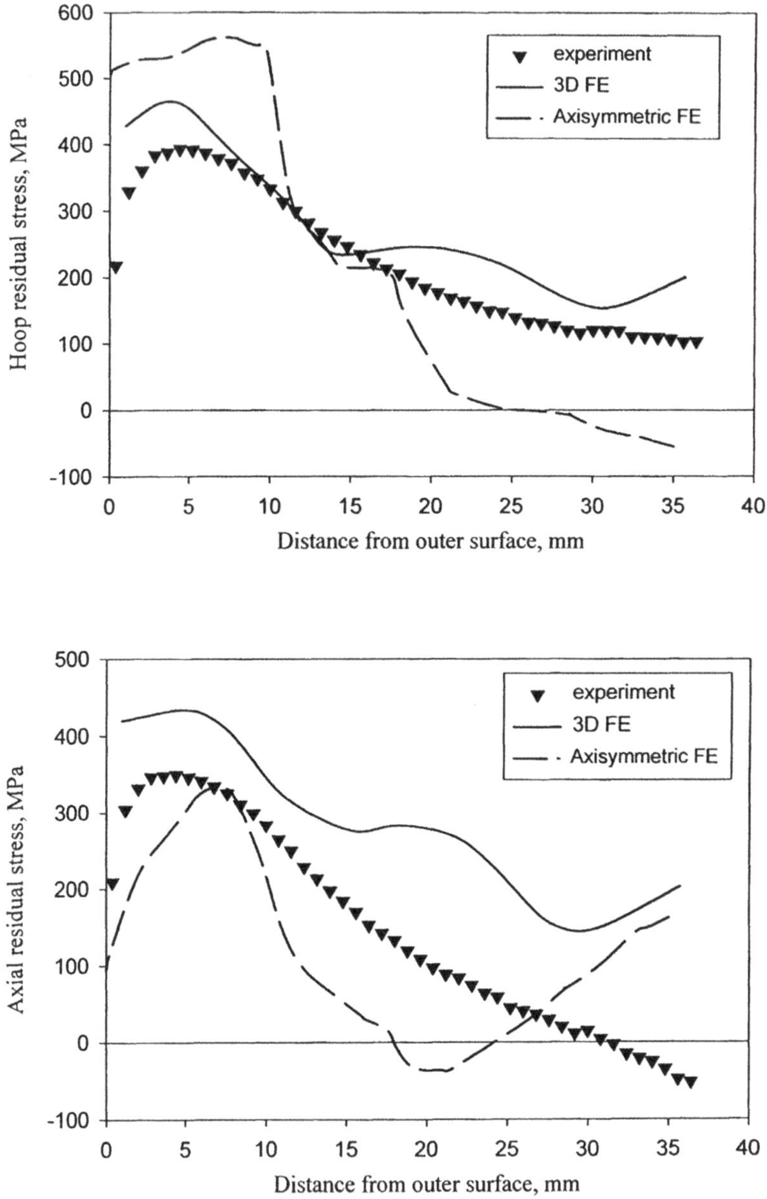

Figure 7 : Residual stress distribution in HAZ with 75% weld repair.

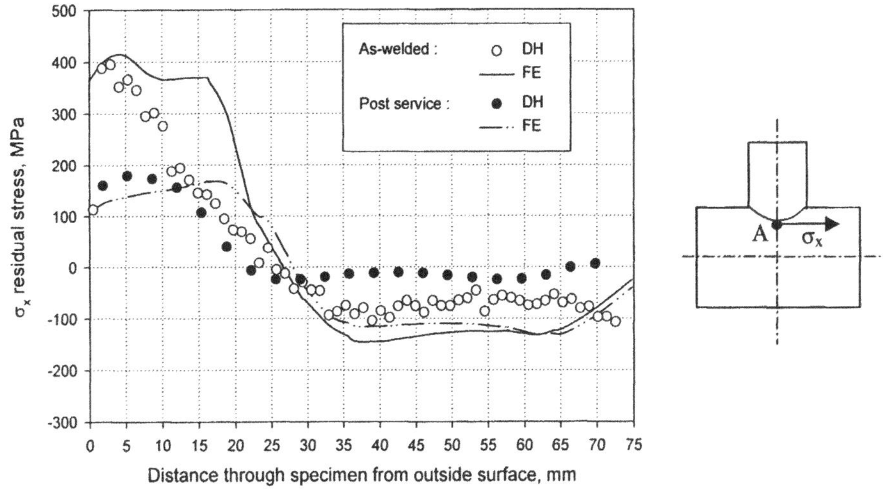

8a) σ_x residual stress

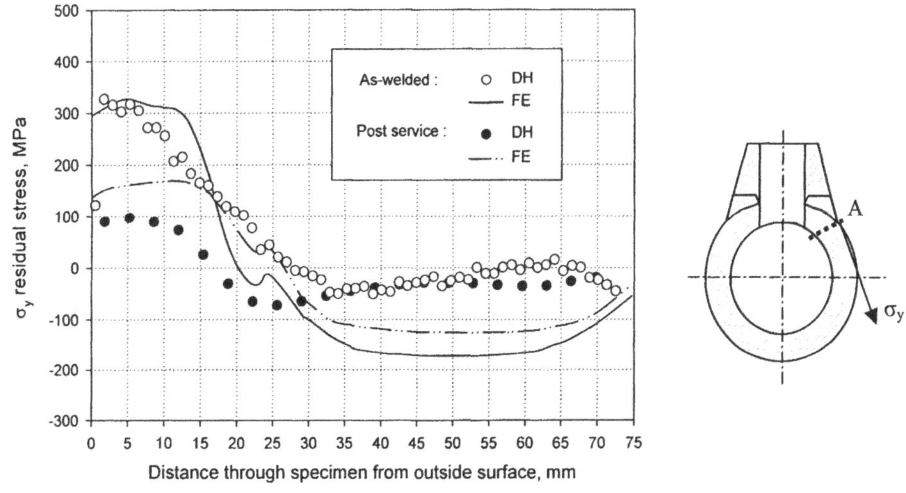

8b) σ_y residual stress

Figure 8 : Variation of measured and predicted residual stresses in a cylinder-to-nozzle component

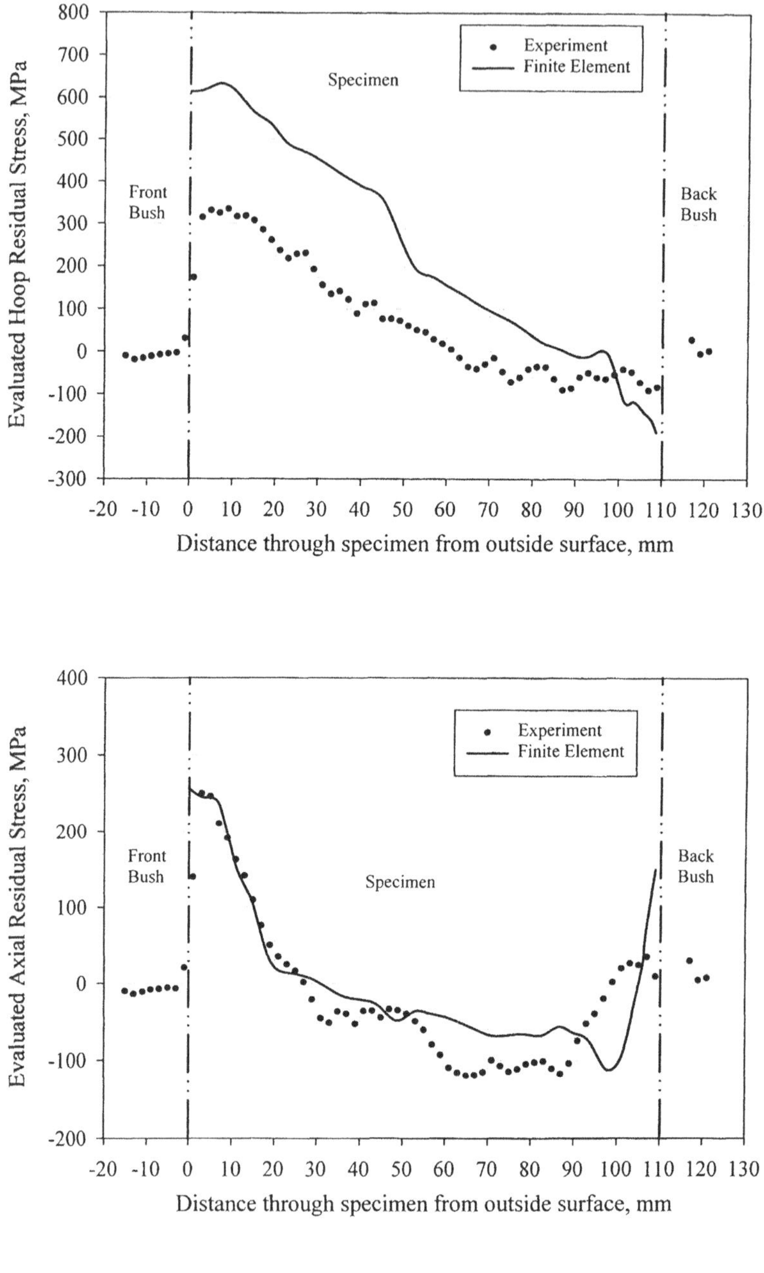

Figure 9 : Comparison of experimental and FE results through centre line of weld of End cap-to-cylinder component

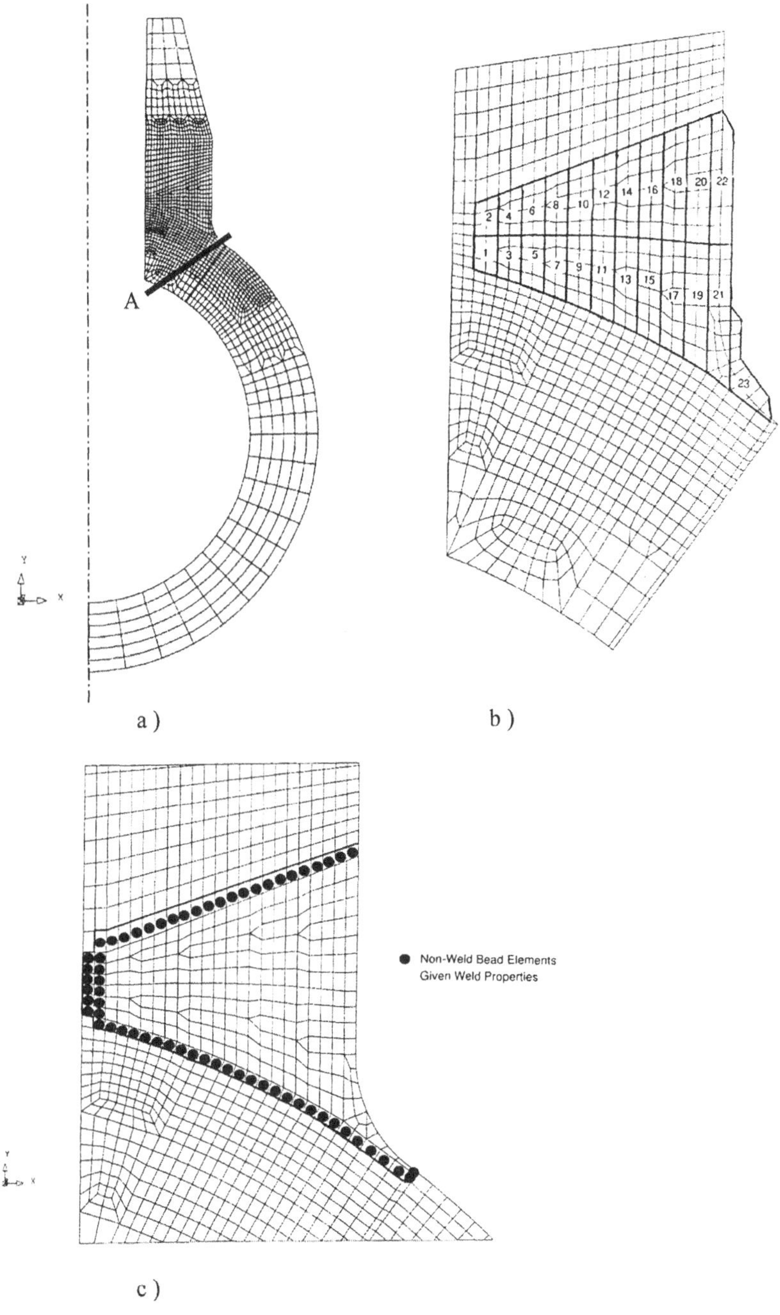

Figure 10 : Finite Element mesh of idealised cylinder-to-nozzle component

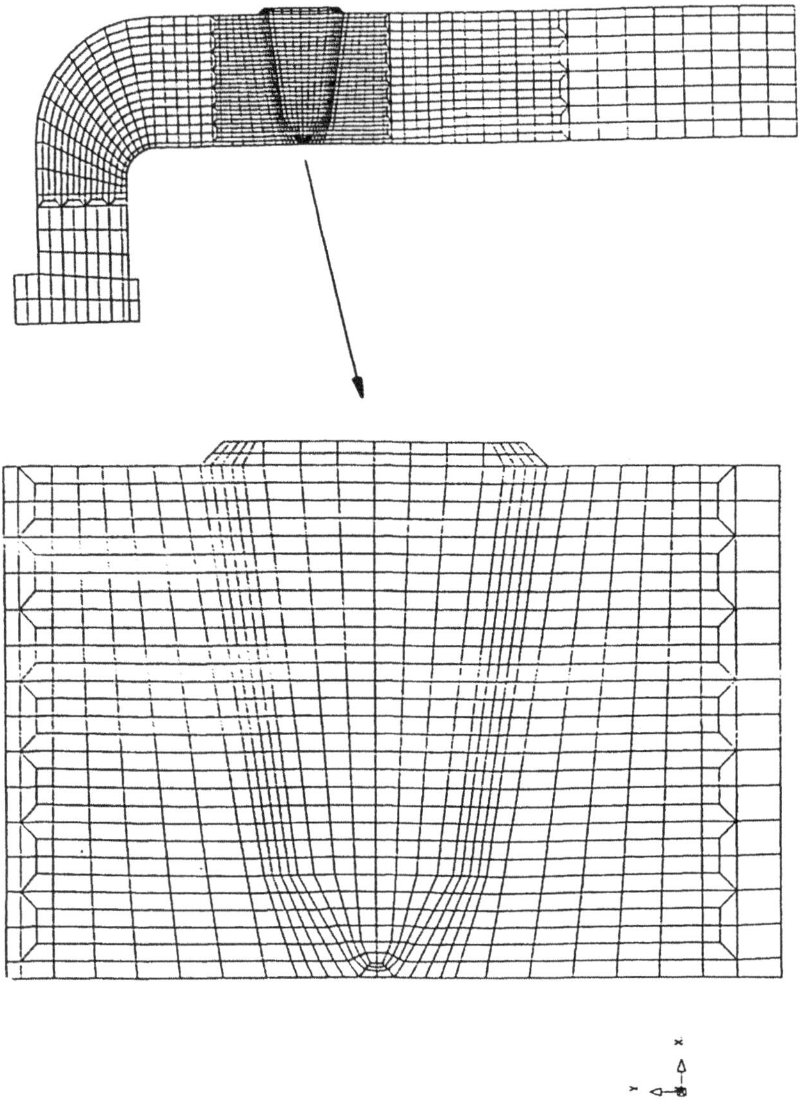

Figure 11 : Finite Element Model of End Cap-to-Cylinder component.

S708/006/2000

Friction stir welding of aluminium extrusions

L D OOSTERKAMP and **O J EMMERHOFF**
R & D Materials Technology, Karmøy and Hydro Aluminium as, Norway
O T MIDLING
Hydro Marine Aluminium, Kopervik, Norway

SYNOPSIS

Friction stir welding is a new welding technique invented in 1992, which has already been accepted for a wide range of applications. It involves plunging of a specially profiled pin into the joint line of two pieces to be welded. Pin is shouldered, so the shoulder is in an intimate contact with the upper surfaces of the workpieces, creating the frictional heat during rotation. Heat induced brings the material locally into a plasticised condition. Motion of the pin shears the surfaces of the workpieces to be welded to the point where solid state bonding can occur. Tool translates along the joint line creating the solid state seam weld. The finite element program FIDAP has been employed for simulating the FSW process. The velocities at the interface between the aluminium plate and the tool are specified assuming full sticking.

1. INTRODUCTION

The friction stir welding (FSW) process has been invented in early nineties by TWI [1] and since then has further been developed and implemented in prefabrication of the ship panels for fast catamarans. The concept is based on the idea of producing large series of panels from relatively small sized extrusions, extruded on medium press capacity (3200 tons). These presses offer good competitiveness comparing to large extrusion presses with respect to price and the delivery time [2]. Extruded profiles are designed in such a way to replace the conventional MIG welding of stiffeners on the rolled plates and the only fusion welding applied is for welding the frames between the panels.
The friction stir welding process has advantages over the conventional welding techniques. Post weld distortion is low indicating lower level of residual stresses. The process is environmentally friendly, welded sections have good surface appearance and mechanical properties.
Friction stir welding has became increasingly attractive since fusion welding of precipitation strengthened alloys, like AA6082 results in significant loss of mechanical properties and many higher strength alloys of 7xxx series are not recommended for fusion welding at all.

2. FRICTION STIR WELDING (FSW) PROCESS

The details of the FSW process are given elsewhere [1-3]. The process involves plunging of the specially profiled tool pin into the joint line of extrusions, which are brought into contact and clamped. The pin is shouldered so the shoulder is in the intimate contact with the upper surfaces of the extrusions. During operation the download force is applied on the tool which rotates and the frictional heat created by this rubbing motion brings the material into the plasticised state. As the tool translates along the joint line, the material is sheared and forced to the trailing edge of the pin where it cools down and consolidates to create the solid state bond.

The main parameters of the process are the download force, revolution and translation speed of the tool, (welding travel speed) and these are the function of the material to be welded (aluminium alloy type and the condition) and the thickness. For each alloy and the temper condition in case of heat treatable aluminium alloys, combination of the welding parameters defines the process window. Welding travel speed (WTS) is a direct measure of the productivity, the larger the speed the higher the productivity. On the other hand, the quality of the weld is to a large extent controlled by WTS and the heat generation that is input via the tool. At a specific power, large welding speed results in cold-lap welding and brittle fracture whereas too low speed at the same power gives local melting and up-set generation.

2.1. Heat Generation During Friction Stir Welding

The heat generated during the FSW process is produced by friction at the interface between the material and the tool. Most of the heat is generated by the rubbing motion of the tool shoulder and much less by the pin. The role of the pin is to shear the contacting surfaces of the extrusions together to the point when these are brought to the inter-atomic distance and the solid state bonding can occur. This process can be compared to the extrusion of the hollow sections using a porthole die, where two separated streams of the material are sheared together to create the extrusion seam weld. [4].

The heat input is controlled by the process parameters and they should be selected such that no melting of the material is caused. Operational temperature of the FSW is somewhere between the eutectic and solvus temperatures of the material to be welded. Too high heat input causes the local melting, resulting in loss of the mechanical properties and too low heat input results in the cold weld.

There are many commonly accepted friction based theories of adhesion between asperities at the contact surface. Since large download force is applied during FSW, it is reasonable to assume that when two surfaces slide along each other under external tangential force, the weld must be formed by a shearing force, which is equal to the friction force. If Coulomb friction model is used, the friction coefficient can be defined as the ratio between the shear and the normal stresses at the contact surface. The coefficient of friction is changing continuously during the welding cycle, from $\mu=1$ at the dry sliding start at the beginning of the FSW, towards zero when the temperature is reached for plastication or asperity melting during the steady state stage. It is, therefore, reasonable to express the friction coefficient in terms of an efficiency friction factor. Experience has shown that friction efficiency factor of $\eta=0.2$ may be used for calculation of the average heat input per unit area and time during FSW. The power generation can thus, be expressed as a product of the friction efficiency factor download force and the maximum angular velocity of the tool (which corresponds to the peripheral velocity at the shoulder edge):

$$q = \eta \, F v_{max} \qquad [W]$$

which is analogue to the expression of power generation used in conventional fusion welding, i.e.

$$q = \eta' \, UI \qquad [W]$$

This expression provides an estimate of the heat input, which may be later used for global structural modelling of distortions of large panels, where detailed modelling of deformation would require high computational resources.

3. MODELLING OF THE FRICTION STIR WELDING

The commercially available program FIDAP [5] has been employed for the simulations of Friction Stir Welding of AA6082. FIDAP is a Finite Element Method (FEM) program based on a Eulerian approach, which calculates the metal flow, stress and strain rates and temperature distribution in the model. The actual geometry of the model is discretised into a large number of small elements (bricks). Material properties and conditions on the boundaries of the model, such as external stresses, velocity and temperature constraints, are prescribed. The governing equations that apply to this problem, i.e. the conservation of mass, momentum and energy, are then solved for all elements in order to obtain the flow field, temperature and stresses everywhere in the model.

The present model is shown schematically in Figure 1. In order to reduce modelling and computational times, the actual geometry has been simplified, so it consists of several entities within the same body. Therefore no contact was simulated. The aluminium and backing plates are described by rectangles surrounding the welding area. The welding tool is cut at some distance above the plate, as the area around the tool is of primary interest for FSW. The tool and the backing devices are modelled to obtain correct temperature in the aluminium, in order to simulate the substantial heat flow out of the aluminium.

The defined geometry is composed of a large number of brick elements, with node points at the vertices of the elements, where the velocities, stresses and temperatures are calculated. The size of the elements depends on the demanded accuracy and the computer resources that are available. Smaller elements are used where the curvature of the geometry is high and where high gradients of the physical quantities are expected. The smallest elements are distributed near the interface of the tool and the aluminium, due to the high gradients of strain and temperature in this area. The current simulations were performed with about 10.000 elements.

3.1. Material Data

Material data and boundary conditions are the crucial for FEM computations. In general the material data for thermally coupled FEM computations of plastic forming can be classified into the thermal data, describing the heat balance in the material and the constitutive equations describing the material response to the exerted load. The latter formulate the flow stress for different strains, strain rates, temperatures and material conditions.

Thermal data are specified by the thermal conductivity λ, which describes the material's ability to conduct heat, and the specific heat c_p, which represents the resistance to change of temperature. The governing equation for the energy balance is given by [5]:

$$\rho c_p\left(\frac{\partial T}{\partial t}+u_i T_{,i}\right)=(\lambda T_{,i})_{,i}+H,$$

where: ρ is the material density, T is the temperature, t is the time, u_i is the velocity of the material in direction i, H is the general heat generation term and the subscript '$_{,i}$' denotes differentiation with respect to a direction x_i. The heat generation term H is zero where there is no deformation and largest near the tool where large deformation occurs. The heat that is generated near the tool propagates by different mechanisms according to the equation for the energy balance: by conduction (λ) by convection ($u_i T_{,i}$) and by diffusion ($\partial T/\partial t$).

The constitutive equations are specified via viscosity as in a Newtonian fluid, by the equation,

$$\tau_{ij} = 2\nu\dot{\varepsilon}_{ij},$$

where the viscosity $\nu = \nu(\dot{\varepsilon}_{ij}, T)$ is a function of the strain rate tensor $\dot{\varepsilon}_{ij}$ and the temperature T. The viscosity is calculated by the Zehner-Hollomon equations, which use a set of material parameters obtained from experimental torsion tests. The tests are performed at low strain rates and the results are extrapolated to the high strain rates that are needed in FSW. There are some uncertainty related to the extrapolation and the method of obtaining the data from the experiments, which are based on average values of strains in the sample and not the local values. This applies in particular to the thin layer under the tool, where very high strain rates occur. The material behaviour in this layer strongly influences the stress, strains and temperatures in the remaining model, so proper selection of the parameters for the material model is crucial for successful prediction of temperature and deformation.

3.2. Boundary Conditions

The boundary conditions in FIDAP can be specified by prescribing stresses or velocities, for the transfer of momentum, and temperature or heat flux, for the transfer of energy. The stress and velocity conditions are relevant on the interfaces between the tool and aluminium and between the backing and aluminium under the pin. In the current simulations, the velocities at the interface between the aluminium plate and the tool are specified so that the layer of aluminium adjacent to the tool is forced to rotate with the tool. In other words, full adhesion between the tool and aluminium is assumed and the velocity of the aluminium surface layer equals the velocity of the tool as shown in Figure 2. This assumption is equivalent to a large down load force where the friction is large enough to make the aluminium stick to the tool. The full adhesion model gives larger deformation and temperature in the material than real stress conditions would have caused, whereas no friction underpredicts the deformation and temperature distribution in the material. Different download forces can be simulated by changing the rotational speed of the tool, allowing the studies of different heat power q applied to the tool, when only velocity conditions are used.

The temperatures calculated by the code depend on the amount of heat that flows out of the model. It is therefore important to specify the temperature or heat transfer coefficients on the boundaries of the model. Heat transfer through the boundaries is modelled as:

$$q = h_c(T - T_R)$$

where: h_c is the heat transfer coefficient, T is the temperature on the boundary and T_R is a reference temperature of the material outside the model. The heat transfer coefficients depend on the outside material and is large for materials with high thermal conductivity. The reference temperature is strongly affected by the heat flux out of the model. The present simulations were performed by specifying room temperature on top of the tool, on the sides of the aluminium and backing, and on the bottom of the backing. The remaining external surfaces were modelled without temperature constraints. The room temperature assumption overpredicts the heat flux out of the model, since the boundaries in the physical model are heated up.

4. RESULTS OF SIMULATIONS AND DISCUSSION

The simulations were performed using typical FSW process parameters and dimensions of the tool and the welded plates. Two different welding speeds were selected: WTS=0.6m/min and WTS=1.2m/min. The first one is slightly less than currently used in the production.
The results of the FE modelling show a strong dependence of the temperature gradients on the welding speed, as shown in Figure 3. For low welding speed, the isotherms are circular, centred at the centre of the tool. At higher welding speed, a substantial part of the heat is carried away due to the translation of the tool, which results in lower temperature in the weld. The isotherms are nearly elliptic. This change of the shape of the isotherms for higher welding speed is attributed to the convective term $U_i\,T_{,i}$ in the energy equation.

It is interesting to observe that the temperature beneath the tool varies significantly along the perimeter as well as in the radial direction. Figure 4. illustrates this distribution at the interface between the tool and aluminium. The highest temperatures occur on the side where the angular tool velocity has the same direction as WTS, i.e. on the advancing side. This is not surprising, since the heat generation increases with increasing shear of the material.
The observation is very important, since weld defects, related to the insufficient heat input are more likely to occur on the side of the weld where the angular tool velocity opposes the WTS, i.e. on the retreating side.

As friction stir welding is a high strain rate process, the strain rate effects are an important issue [6]. The 3D-color plot of the shear strain rate is shown in Figure 5. This plot gives an indication where the highest temperatures are expected. The highest shear strain rates are observed near the edge of the shoulder and in the root at the end of the pin.

As already said, the quality of the weld is influenced by the proper selection of the process parameters and the productivity is determined by the WTS. To be able to increase this parameter, without compromising the weld quality, the heat loss at the root to the backing device had to be minimised. FE simulations were run for two different materials of backing devices: steel and ceramic. The backing plates that are currently used are made of steel, which has a relatively high thermal conductivity. As such they are efficient media for transport of heat. This results in large temperature gradients away from the shoulder towards the root of the weld, which may result in poor mechanical properties at the root of the weld, leading to cold welds. Using a backing with low thermal conductivity, i.e. with the properties of a heat insulator, an increase of the temperature in the root of the weld is expected.

The FE simulations have demonstrated that in case of the ceramic backing, the heat dissipation at the aluminium-backing interface is significantly lower and therefore the temperature gradients are smaller. This results in relatively higher temperatures in the root of the weld. The simulations with steel backing were performed with full adhesion between the aluminium and the backing, whereas the cases with ceramic backing were performed without friction on the backing. With more realistic conditions for both cases, i.e. less deformation when using the steel backing and more deformation when using the ceramic backing, the difference in weld root temperature will be even higher than in the present simulations.

5. CONCLUDING REMARKS

A Finite Element code has been used for simulations of temperature, deformation and stress distribution in FSW process. The thermal and mechanical data for the different materials are included in the model. The governing equations are solved for the mass, momentum and energy balances. The aluminium is assumed to adhere to the tool, thus neglecting the frictional conditions between the tool and aluminium. Although the calculated temperature field is overpredicted, the model is an excellent tool for optimisation of process parameters. Since it utilises Eulerian approach it can not be used to predict the residual stresses. However, the deformation of the material during FSW can be predicted. Experience has shown that the tool pin design strongly influences the weld quality, and the deformation gradients are different for different pin designs. If the friction between the tool and the material is included, the model can serve as a powerful and cheap device for FSW process and tool optimisation.

The use of different backing material showed the largest influence on the temperature distributions in the welded material. Using a ceramic backing leads to higher temperatures in the root of the weld and more even temperature distribution across the welded plate than using a steel backing. This allows for higher welding speeds before cold welds and the resulting brittle fracture of the weld root would occur.

6. ACKNOWLEDGEMENTS

The authors would like to thank Hydro Aluminium a.s for their permission. For financial support to present this work we would like to thank the management of R&D Materials Technology.

7. REFERENCES

[1] Int. Patent Application, PCT/GB92/02203, *Improvements related to friction welding; Friction stir butt welding,* The Welding Institute, Cambridge, UK, Dec.1992.
[2] O.T. Midling, H. G. Johansen, *Production of wide aluminium profiles by solid-state friction stir welding,* Int. Aluminium Extrusion Technology Seminar & Exposition, ET96, Chicago, Illinois, 14-17 May 1996.
[3] Midling O. T., "Friction Stir Welding – A Valuable Processing Route", Proc. of the Aluminium 97 Conf. Essen, Germany, September 1997.
[4] Akeret R. "Extrusion welds – Quality aspects are now center stage", Proc. of the Extrusion Technology seminar, ET 96, Chicago 1996.

[5] FIDAP Users Manual, Version 7.0. Fluid Dynamics International, 1993.
[6] O.T. Midling, L. Djapic Oosterkamp, J. Bersaas, *Friction stir welding aluminium – Process and applications,* Proc. 7th Int. Conf. on Joints in Aluminium (INALCO98), TWI, Cambridge, UK, April 1998.

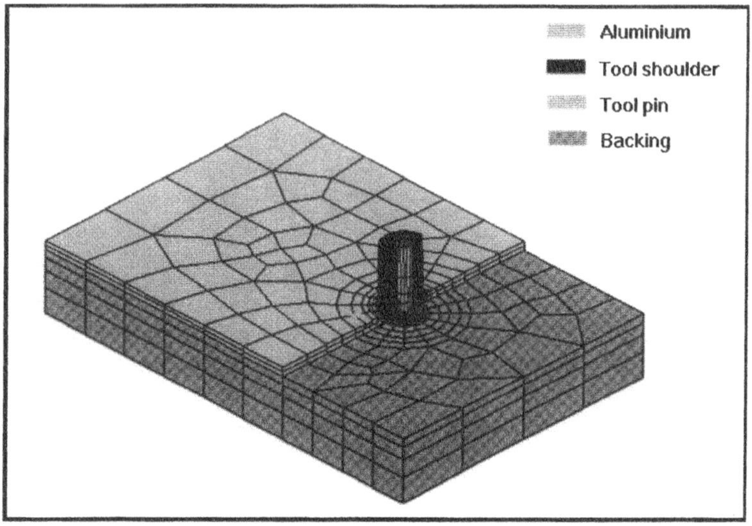

Fig 1. Finite element mesh of FSW model

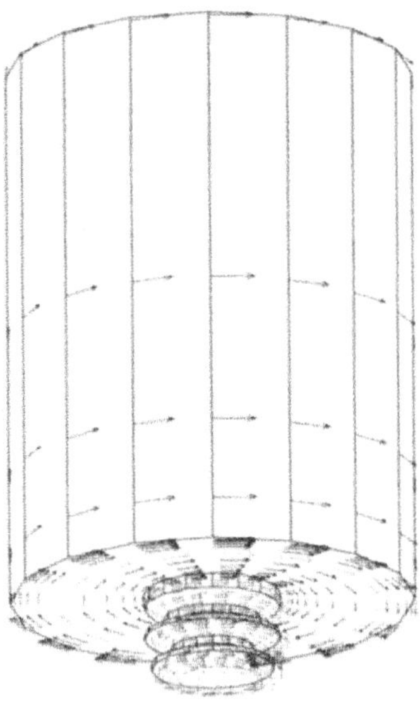

Fig 2. Velocity boundary conditions applied on the tool and the pin. The velocities describe the rigid body rotation of the tool, reference vector on the tool shoulder is 1 m/s

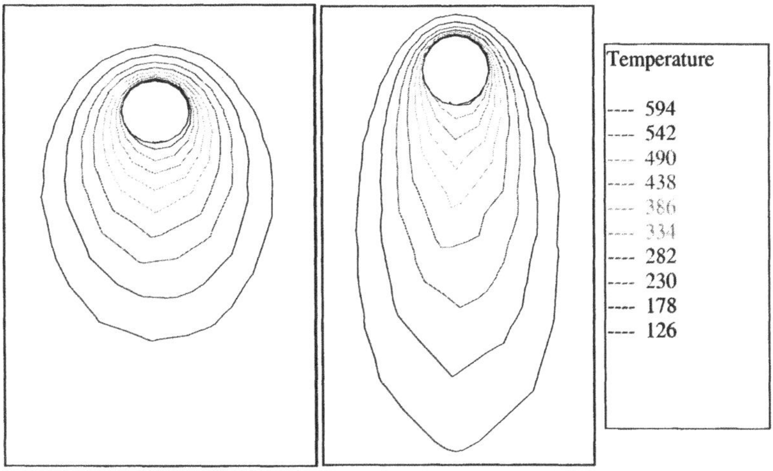

Fig 3. Isotherms on the weld face around the tool. Welding speed WTS=0.6m/min for left plot and WTS=1.2m/min for right plot. Tool diameter: 15mm.

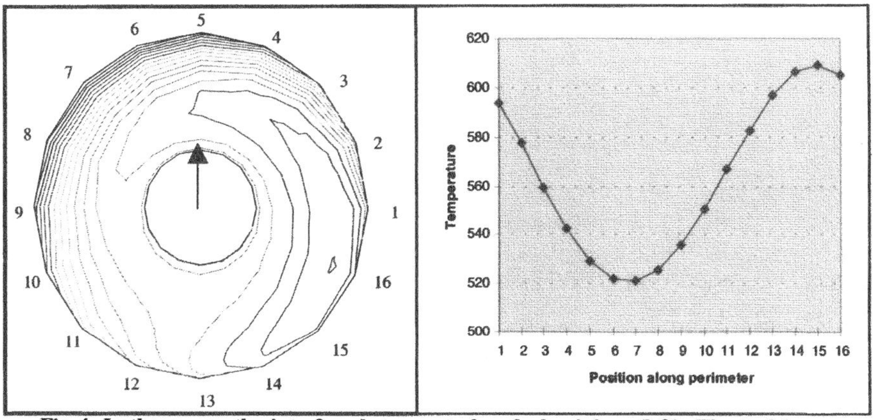

Fig 4. Isotherms on the interface between tool and aluminium (left). The tool is moving as indicated by the arrow, and is rotating in the counter-clockwise direction. The plot on the right shows the temperatures along the perimeter of the tool, as indicated by the numbers 1-16.

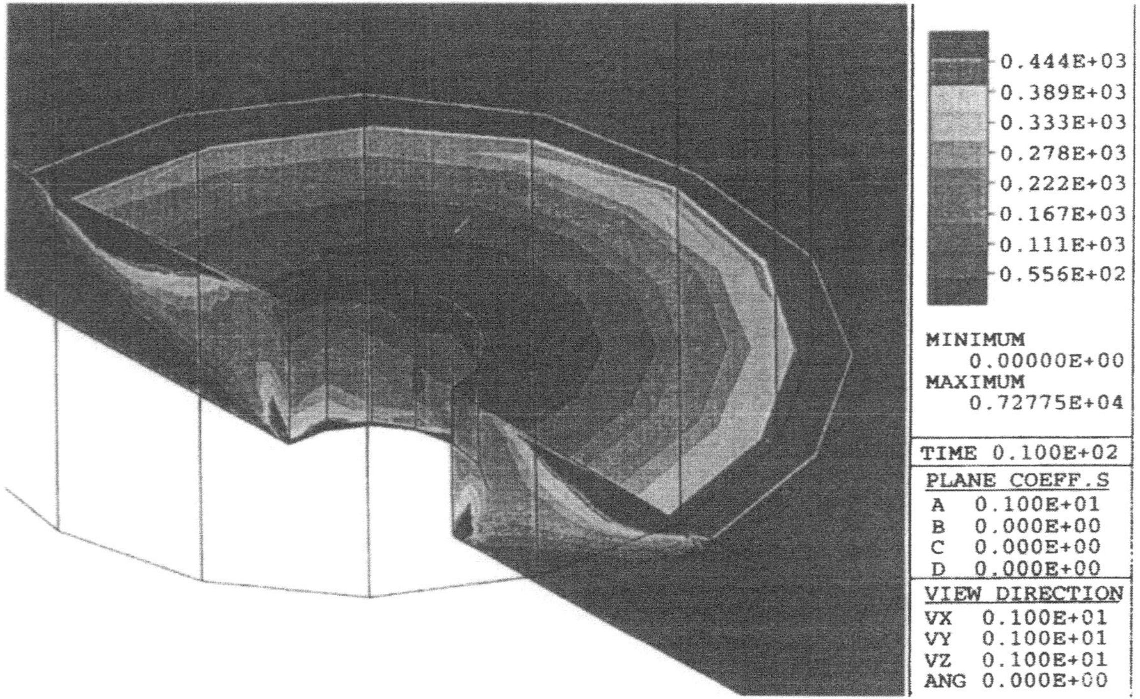

Fig 5. Shear strain rate distribution around the tool shoulder and tool pin during FSW

S708/007/2000

Integrity of welded joints in pressure vessels

R K PENNY
R K Penny & Associates, Noordhoek, South Africa
W KOHLHÖFER
Peninsula Technikon, Bellville, South Africa

1. Foreword

The integrity of welded joints is of prime importance from many viewpoints, not least of which are safety and economy of plant operation. A previous IMechE seminar (1) discussing high temperature welds, indicated that welded joints in pressure vessels are the least tractable in terms of design. Progress since then is minimal. This is particularly true in the power generating and petro-chemical industries in which welded vessels are subjected to aggressive environments and loading whilst being expected to sustain economic lifetimes beyond 30 or more years. In these cases the vessels are often subjected to temperatures within the creep range causing internal cavities to be generated which can be life-limiting at fractions of the design lives. The welded joints at nozzle intersections in these cases are clearly limiting factors in the overall success – both operational and economical – of the industries concerned.

In assessing the integrity of welded joints, several areas of uncertainty arise. The most obvious of these include the weld manufacture itself: both from processing and control, variations in material properties within the weld regions adjacent to the nozzle: as–treated and after heat treatment. Each of these factors carries uncertainties, not only from vessel to vessel but within each vessel. Furthermore, methods of inspection of the welds are incapable of reliable results (2) not only arising from the presently used inspection technologies and operator skills, but also on account of inaccessibility.

Codes of Practice usually seek to overcome these areas of uncertainty by using "safety factors" during the design process; these factors of ignorance are frequently based on past experiences. Such factors can not only lead to un-safety – particularly at regions where extra thicknesses of parent material can give rise to more areas of uncertainty – but, just as importantly, they can lead to unnecessary costs and extra initial costs of plant at the design/manufacture stage which have to be absorbed. Once the plant is in operation it has to be maintained – usually at regular intervals – which can be longer or shorter than necessary – and overall are managed by emergency, to prevent loss of life and product. Rational design, and particularly at points of high risk such as welded joints, cannot be left to the traditional Codes of Practice alone: supplementation through probabilistic techniques is essential. This presentation describes recent efforts in deriving rational methods for welded joint reliability. The results only show early promise at rudimentary stages so far mainly because an essential requirement of the work is that it should be of practical significance to designers rather than demonstrations of academic erudition which, like codes of practice, are already profuse. The main ingredients in the presentation involve the potential for use of SYSWELD® (3) to simulate the welding process and its effect on material properties and their distributions at welded joints; another commercial package ABAQUS® (4) is employed in integrity assessment.

2. Brief descriptions of commercially available FE routines

2.1. SYSWELD®

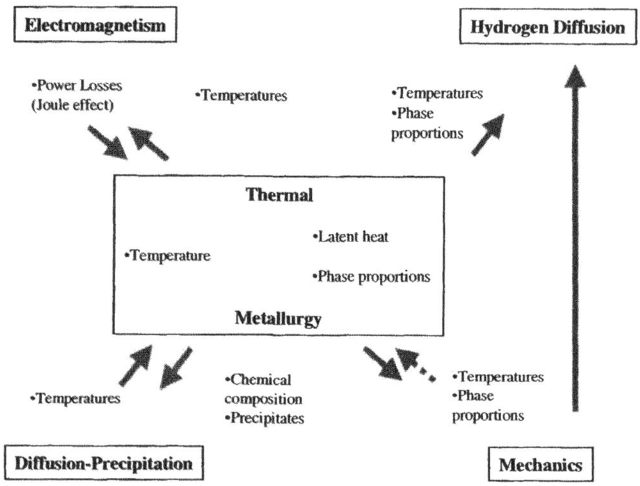

Fig 1. SYSWELD+ General Architecture (3)

Heat treatment and welding processes induce residual stress and strain. These can have a major effect on the mechanical strength of the component. Using numerical simulation as a complementary tool to experimental techniques will lead to a fuller understanding of the interactions between the involved physical phenomena. SYSWELD+® targets production processes such as:

- Quenching
- Tack welding
- Welding with or without feed metal (TIG or MIG)
- Other forms of surface treatment

2.2. ABAQUS®

ABAQUS is well proved program capable of structural analyses involving inputs of various loadings and material properties. It is generally available internationally and is economical in use by practicing engineers.

3. Some experiences so far

3.1. Using SYSWELD®

As simulations alone are meaningless, some work has been performed verifying repair work (5) on steam generators using SYSWELD®. Defects markedly shorten the service life of such plant. As every intervention made has an effect on the service life, it is necessary to know the effects of repairs on such components. One possible way of determining this is through numerical simulation. The entire repair procedure is simulated numerically and the levels of residual stress are determined which are then taken into account when determining the component's service life.

Two models were constructed, one for mechanical and the other for thermal verification. A schematic of the mechanical model is shown in Fig 2a.

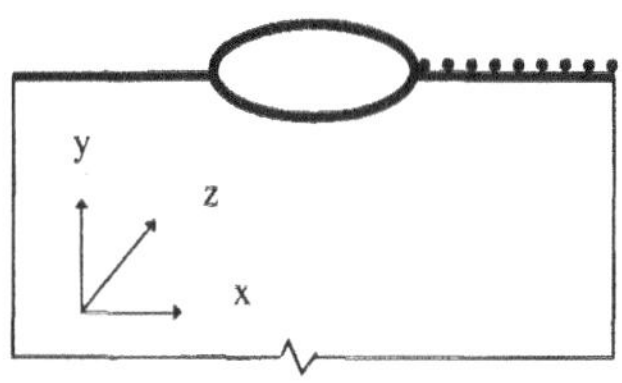

Fig 2a. 2D Model

Fig 2b shows a reasonable correlation between experimental and numerical results. The dotted line on the right hand side of the weld indicates the position where the results were measured. These results may then be used in combination with the external loads – pressure, temperature etc. to calculate fatigue and creep life. The thermal model was used to verify thermophysical properties such as hardness. The heating and cooling rate was observed and the effect on hardness calculated and measured.

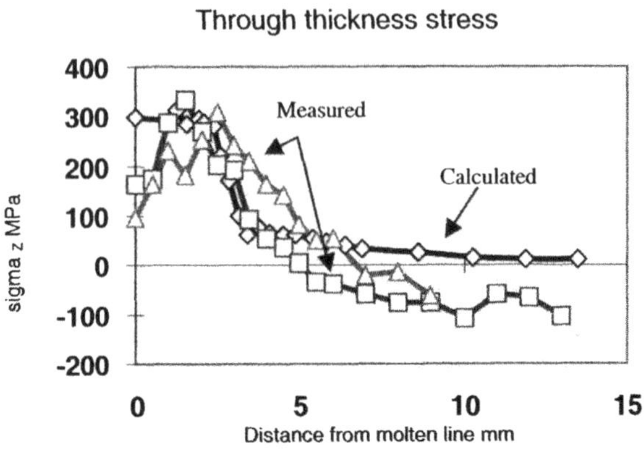

Fig 2b. Residual Stress Comparison

3.2. Using ABAQUS®

It would be useful to the present objectives if a probabilistic finite element package were available. This ideal is illustrated schematically in Fig 3; unfortunately such methods are not commercially available yet.

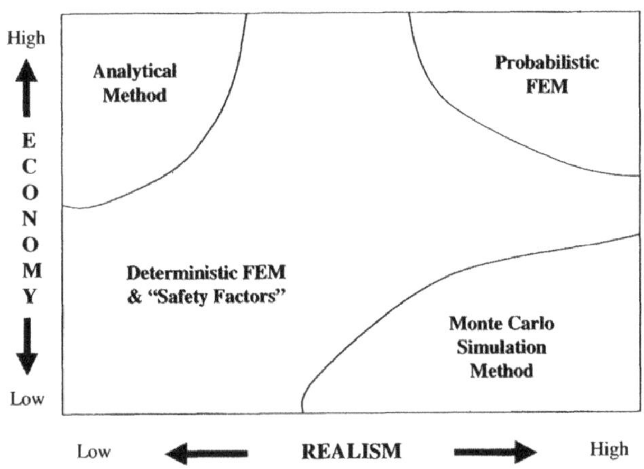

Fig 3. Schematic showing computational procedures for probabilistic methods

So far, the main activities in using ABAQUS® have been with extensive trials on the effects of material property variations at welds in nozzles. Figure 4 shows the nozzle - pressure vessel intersection and the section under investigation.

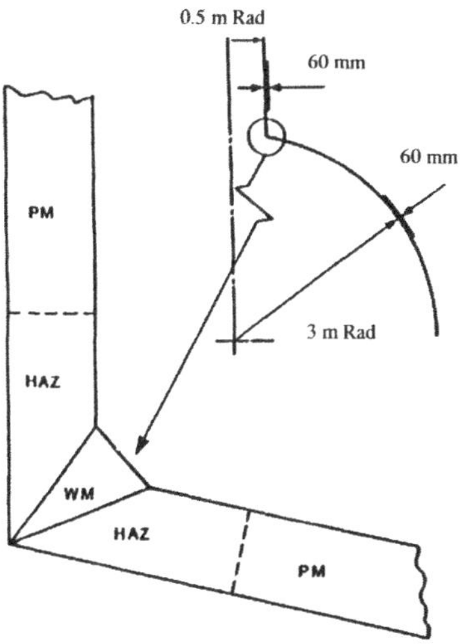

Fig 4 Schematic of model dimensions

Table 1 gives a list of the material properties used in the deterministic as well as the probabilistic cases. The vessel was subjected to an internal pressure of 10.5 MPa and the material used was assumed bi-linear with a yield point of 500 MPa. Figure 5a (7) shows the results obtained using no variation in the material properties and similar characteristics for parent metal, heat affected zone and weld metal. This may be compared to figure 5b where different Young's moduli were used for the WM, HAZ and PM respectively.

Table 1. Material Properties

Weld Metal (WM)	Heat affected Zone (HAZ)	Parent Metal (PM)	SimulationType	Figure
210/0/500/0	210/0/500/0	210/0/500/0	**Deterministic**	5a
256/0/500/0	228/0/500/0	210/0/500/0	**Deterministic**	5b
256/0.10/500/0.20	228/0.05/500/0.05	210/0/500/0	**Probabilistic**	5c
256/0.20/500/0.20	228/0.10/500/0.10	210/0/500/0	**Probabilistic**	5d

This has been accomplished by using the usual procedures for model geometry discretisation into elements but with the assignment of properties to each element in terms of a local value and its coefficient of variation. Typical results are shown (Fig 5c) for welds which are not only variable in properties but can include voids (caused by slag inclusions or lack of fusion) on a randomised basis throughout the weld/HAZ/parent metal region, by using a Monte Carlo process and a random number generator. Figure 5d shows results for an even greater variation in material properties.

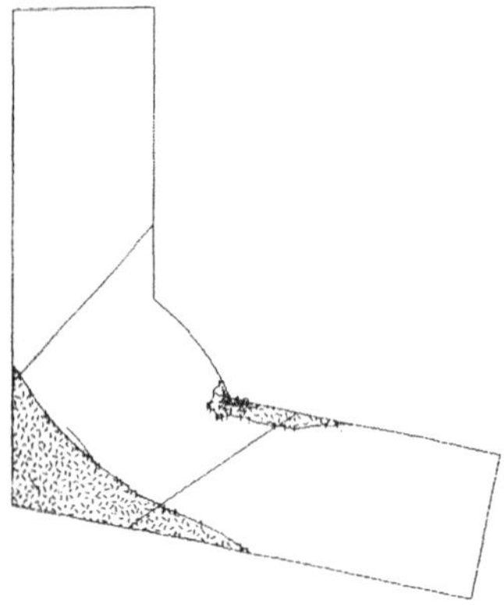

Fig 5 a.

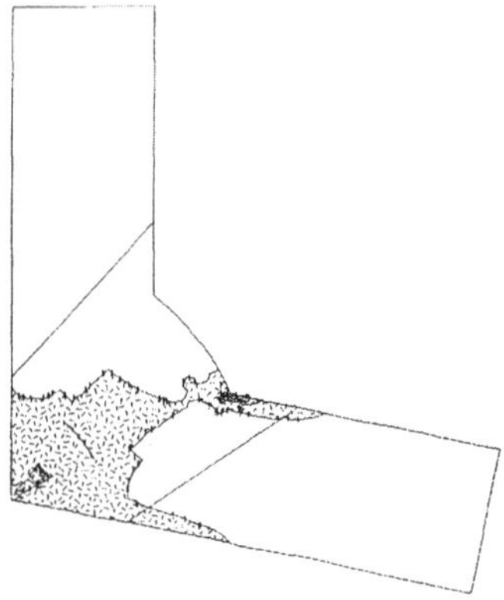

Fig 5 b.

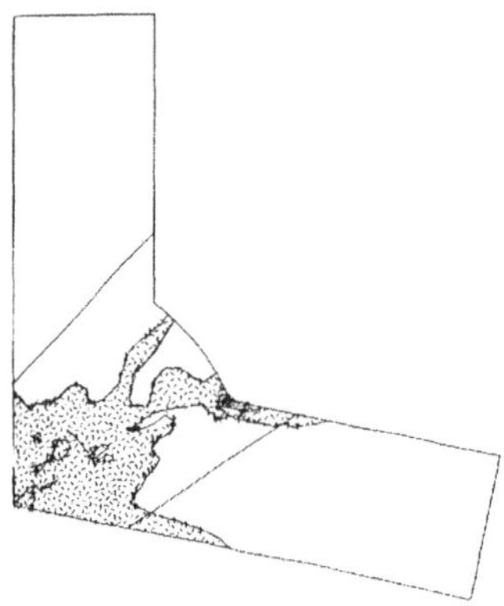

Fig 5 c.

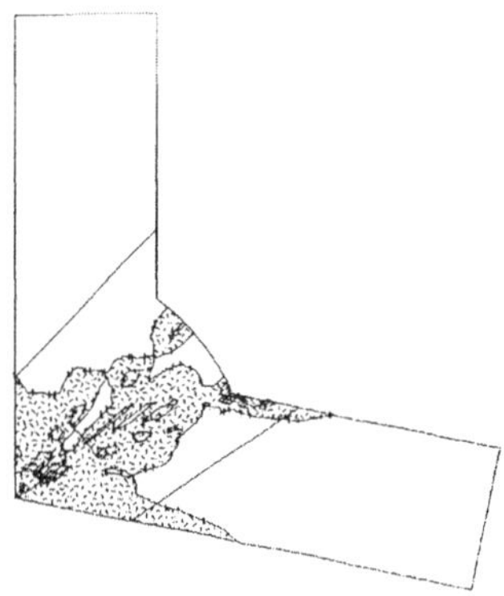

Fig 5 d.

As stated earlier, the purpose here is to establish the principles of using material property randomisation with a readily available computer code. In the case illustrated, material properties have been provided by hardness values established by scanning an available welded joint (6). This type of analysis seems to work well (7) and has been used with ease by students and practitioners with elementary backgrounds in finite element analysis.

Reactor Pressure Vessel (Figure 6)

This nozzle was connected to an inlet pipe and the vessel dome by a welded joint. The vessel had been in operation for almost 200,000h at a nominal temperature of 500°C. Replica treatment in the weld region revealed degradation by creep in the parent metal/heat affected zone on the nozzle side where cavitation and micro-cracks were found. No cracks were reported by ultrasonic investigation.

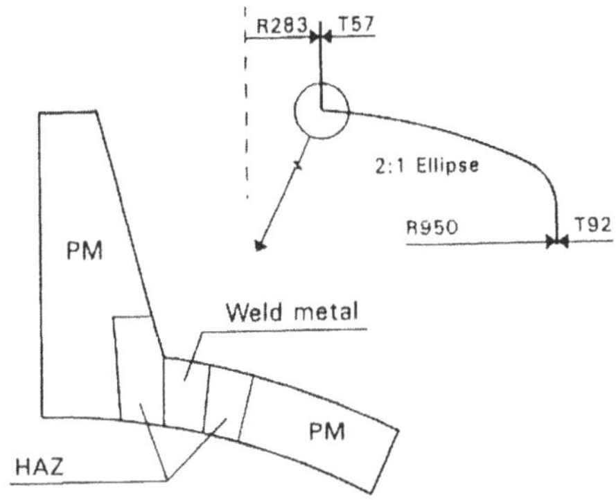

Figure 6. Reactor pressure vessel nozzel

A deterministic creep and fracture analysis which included pressure and piping reactions indicated that the vessel integrity would not be impaired until the next shutdown was due (12 months), provided no extreme pressure/temperature excursion occurred. Of concern was that an extreme load excursion, coupled with weakened strength may lead to sudden and catastrophic failure – by exceeding the current static strength or through a combination of an undetected crack and uncertain critical fracture toughness.

The yield/ultimate strength due to creep is known to reduce rapidly at high life fractions (8), but the extent of damage in the present case, is uncertain because only surface (replica) measurements were made. Likewise, NDE techniques found no cracks, but these techniques are also uncertain. Available data for fracture toughness shows a considerable amount of scatter and varies with temperature and stress. The remaining strength of the nozzle must therefore be considered as a random quantity and the magnitude of the mean strength had to be estimated from available data and the magnitudes of strength variance assigned by engineering judgement and, if necessary, a sensitivity analysis performed.

Sudden failure due to pressure/thermal excursions was likely to be dominated by pressure and nozzle reaction effects. Historical data indicated a coefficient of variance (COV) of about 7%, but an extreme value of 15% was assured for future operation over the next 12 months. Mean piping system reactions were estimated on the basis of a standard system reaction analysis. All problem parameters are summarised in Table 2.

Finite element analyses which included zones of parent metal, heat affected and weld metal zones were performed on the entire model in the manner described earlier. From this mean, standard and reliability induces were evaluated using the data in Table 2. These analyses indicated that the most probable points of high Mises and principal stress occur at the intersection of the heat affected zone and weld metal crotch. Repeated analyses were performed at this most probable area where stress was found to die away to about 60% of maximum values two thicknesses away from the weld.

TABLE 2 - Problem parameters – reactor pressure vessel nozzle

Parameter	Description	Mean	STD	COV
P	Internal pressure (MPa)	3	0.45	15%
M	Piping system reaction moment (kNm)	150	22.5	15%
E	Elastic modulus (GPa) (PM, HAZ & WM)	160	24	15%
σ_y^o	Initial yield stress (MPa) (PM)	250	37.5	15%
σ_y^o	Initial yield stress (MPa) (HAZ & WM)	200	30	15%
H	Hardening modulus (GPa) (PM, HAZ & WM)	1	0.15	15%
σ_{f_s}	Static strength failure criterion (MPa)	350	35	10%
σ_{f_f}	Fracture strength failure criterion (MPa)	300	36	12%

The results indicated a probability of failure less than 1×10^{-4} but the pressure at which this would occur is higher than the shutdown levels monitored by warning systems. The conclusion was that the likelihood of failure was small in spite of the presence of creep damage. Operation could continue but the state of material damage should be intensively examined at the next shutdown before further operation.

4. Observations to date

4.1. With SYSWELD® more valid and appropriate experience is needed in order to ensure that relevant data and material properties in the region of welds can be established with ease and economy. Results obtained need physical validation and accreditation to make the tool reliable, accessible to practitioners and economically viable. To accomplish this it is likely that international cooperation on a well - planned basis is essential. To this end, cooperation between some universities in South Africa and Germany has started.

4.2. Experiences with ABAQUS® have shown that material property randomisation in welds can be assessed without any further development of the program. Results obtained so far with the suggested methods substantiate concerns that welded joint reliability is not likely to be tackled in the usual way through deterministic methods. It is essential therefore that the simulations which could be provided with programs such as SYSWELD® be linked with the approach described here using ABAQUS® through Monte Carlo techniques. It is also anticipated that further synergies between the use of SYSTUS® systems and ABAQUS® will be built upon for high temperature design based on continuum damage mechanics (9), (10).

5. Conclusions

- Structural methods for assessing effects of material randomness at welded joints can and should be made readily available through standard FE programs. Engineers need to be encouraged to use these methods to evaluate probabilities of failure.
- Material inhomogeneities in welded joints can be simulated but further work is essential if the methods presently available are to be of practical and economic use in reliability assessments.
- Achievement of everyday use of the principles outlined here will be made sooner and more profitably to industry if international collaboration is furthered.

References

1. "Life of welds at High Temperatures," IMechE Seminar, 2 – 13 June 1990, London.
2. P Kauppinen, J Sillanpaa, "Reliability of Surface Inspection Techniques." Int. J Pres V & Piping, 54, 1993.
3. "SYSWELD, Simulation of the Effect of Welding" (users manual, revision C, Framasoft+CSI, 1994).
4. ABAQUS Users Manual, version 4.5, 1984 Hibbit, Karlsson and Sorenson Inc., Providence, Rhode Island.
5. L Junek et al., The Effect of Repair Welding Residual Stress on Steam Generator Lifetime, PVP – Vol. 373, Fatigue, Fracture, and Residual Stress 1998, Page 377, San Diego.
6. W Kohlhöfer & R K Penny, "Dynamic Hardness Testing of Metals." Int. J. Pres. Ves. & Piping 61. (1995) 65 – 75, Elsevier Science Limited
7. R K Penny, Simulation of Effects of Material Randomness on Structural Behavior, Proc. 6th World Conference, Structural Analysis and Optimisation, FEMCAD, IITT-International, Paris, 1989.
8. Penny, R.K. The Usefulness of Engineering Damage Parameters During Creep. J. Metals and Materials, vol. 8, 1974.
9. SYSTUS (FRACTURE MECHANICS) Ref 24. J Lemaitre, Unified damage laws formulation, C.R. Acad. Sciences, Paris, T.305, serie II, 1987.
10. R K Penny, DL Marriott, "Design for Creep." 2nd edition, 1995 Chapman & Hall.

S708/008/2000

A study of finite element meshing methods to be used in welding simulation for ship structures

A C RAMSAY
MSC Software Limited,Camberley, UK

Introduction

This work was initiated to support a project to determine the residual stresses and deformations in ship structures resulting from thermal effects from the welding process. The project aims to develop predictive methods of assessing the effects of the welding process using the Finite Element Method and validate these methods with reference to experimental data. The study here is was initiated as a result of doubt placed on the accuracy of initial models of the 3-dimensional welded assembly, primarily due to the occurrence of very high values of principal stresses on the interface between the parent and weld metals. The need to conduct a complex 3 dimensional simulation that is highly non-linear will demand large computing resources and this study will also provide information to help decide on mesh densities which are feasible and sufficiently accurate.

1.0 Effects to be studied

The complete process of weld deposition including thermal, mechanical and metallurgical events is an extremely complex phenomenon. This study concentrates on the thermal and mechanical behaviour, and in the main ignores metallurgical events as affects both the thermal and mechanical behaviour. There is no justification for this other than a desire to examine general characteristics of the simulation methods. It is intended to later examine the effects of microstructural changes on the thermo/mechanical behaviour both through solidification and later re-crystallisation. The latent heat of solidification has therefore been ignored and the properties of the weld metal prior to solidification treated as a soft solid, by extending the lower temperature thermo/mechanical behaviour into the liquidus temperature range. Some sensitivity studies were done on these assumptions, and found that the final stress distributions on cooling were not much affected by this assumption. This does not infer, however that ignoring the true metallurgical behaviour of the material is similarly unimportant. It is believed that inclusion of these effects is important to the accurate simulation of the welding process, but the focus of this study is on developing valid finite element models to give an adequate representation of the thermo/mechanical behaviour. When this is achieved, there will exist an accurate modelling procedure where the thermal and strain effects of metallurgical processes can then be adequately simulated.

The simplified process that this study seeks to simulate is therefore
1. start with a cold plate with the weld metal absent
2. pre-heat the edge of the plate on the assumption that the welding process brings the temperature of the plate edge surface up to iquidus during the deposition of the weld
3. introduce the weld metal as liquid at an appropriate temperature, mechanically connected to the plate
4. Allow the composite component to cool.

The target process for this technique is a multi-pass process, but the methodology being developed is valid. Further deposition is a repeat of the simulation with different initial mechanical and thermal conditions, retained as an output of the previous deposition.

2.0 The finite element model

The structure to be studied is very simple and illustrated in figure 1. It is a simple butt weld of two plates with an idealised squared off end to the plates. It is assumed that the weld is laid down from both sides allowing a quarter of the section of figure 1 to be represented by a finite element model in figure 2. The welding process is obviously a tri-dimensional event, but for the purpose of studying the effectiveness of modelling methods, it is represented by a two dimensional section of the structure, with a nominal thickness. This assumption is reasonable in that the heat transfer out of the plane of our sample slice, heating the plate prior to deposit will be balanced in a steady process by the preheating of this section prior to deposition. What will be affected is the thermal history of the process which will be changed by approximating the pre-heating and underestimating the post-cooling rate. Of more significance is the assumption of plane strain (no strain out of the plane of the problem) which may overestimate the hydrostatic stresses developed in the model.

A valid analysis of a real welding process must in the end be a tri-axial simulation, but the general modelling principles developed here will remain valid.

The modelling options considered are as follows
1. Mesh density
2. Element class (lower or higher order elements)
3. Element integration scheme (order of variation of element strain)
4. Assumed temperature data or coupled thermo-mechanical analysis
5. Pre-heat mechanism
6. Use of inactive/active elements
7. Use of Contact algorithm

3.0 Initial conditions

The simplest approach to this problem is to assume some sort of changing temperature distribution and carry out a simulation of the mechanical behaviour considering the

temperatures as a form of loading producing thermal strain. When carrying out mechanical analysis with temperature distributions, the temperature is not calculated, but imposed, and these temperatures in the MARC program are regarded as properties of the elements and defined as state variables. There are therefore no variables at the nodes describing temperature. This has an advantage in that the starting condition for the simulation can have elements with different temperatures connected to the same nodes. The idealised case of attaching hot weld to cold base metal can therefore be investigated.
The general approach to the problem is to carry out a coupled analysis where the nodes of the finite element model have, in addition to the mechanical displacement degrees of freedom, a thermal degree of freedom which allows the algorithm to calculate the changing temperatures with time. In this case, the initial temperatures must be defined at the nodes of the problem, which for continuity implies that adjoining elements have the same temperature at the junction. Any idealised study, which requires a discontinuous element temperature field as an initial condition, is therefore inappropriate with this method. It is much more suited to the 'real' problem where a complex system of heat transfers is occurring before, during and after the deposition of the weld. The accurate representation of the mechanical behaviour then depends on a realistic temperature history throughout the model. In this case the initial conditions will be a constant cold temperature of the pate and a similarly constant, hot temperature in the weld metal, but the difference is that these parts of the model will be initially disjoint and a modelling device will be used to join them. There are two standard methods of joining meshes available. One is to make elements inactive until a particular time and then switch them on when required. The problem of shared nodes in a coupled analysis is extant here and the new elements will have the existing nodal temperature imposed on their common nodes. Another is to use a contact algorithm to couple the meshes together, where each has their own boundary nodes and a heat transfer coefficient can be defined to control transfer of heat from one part of the model to another.

4.0 Mechanically only model and element selection

In the MARC program there is a wide range of elements for many types of simulation. In this application we are restricted to four types of plane strain elements as follows.

Type 1. Lower order 4 node with 4 integration points and linear distribution of internal strain.

Type 2. Lower order 4 node with 1 integration point and constant distribution of internal strain.

Type 3. Higher order 8 node with 9 integration points and quadratic distribution of some internal strain components.

Type 4. Higher order 8 node with 4 integration points and linear distribution of some internal strain components.

The effectiveness of these elements are examined using a simple mechanical model where the weld part starts at an initial temperature of 1500 degrees centigrade and the parent metal at 20. This is defined as a stress free state. The weld metal is then cooled to 20 deg. C and the resulting stresses examined. There are two ways of looking at stress. Element stresses that reflect the strain calculated at the integration points, and nodal stresses, where the element stresses are extrapolated to the nodes and the values from the joining elements then averaged. The second is the more usual method of displaying FEA results in that the distributions appear relatively smooth and continuous; it is valid as a representation of the continuous strain field that exists in a component. The first method produces a discontinuous distribution which is useful for diagnosis of results quality, but does artificially increase local unaveraged nodal values and can lead to an over estimate of peak values. Figure 3 shows the averaged and unaveraged results for the 4 types of element on a basic mesh refinement of the model. Table 1 shows the peak values predicted in the elements for those models. A comparison of these results indicates that for this discontinuous temperature case, the fully integrated elements (types 1 and 3) exhibit a strain distribution caused by the compatibility forces, which exaggerates the local element strain distribution. This was the effect seen in larger models that prompted this study. In the general modelling of welding problems, it is likely that discontinuous temperature field will be a common occurrence, so it is recommended that only reduced integration elements, of whatever order desired, are used.

The use of simple type 1 elements is normally avoided in modern usage. The fact that only a single stress result is calculated and subsequently transferred (not extrapolated) to the nodes ensures that the surface values reported are likely to be an underestimate of the true surface stresses. This disadvantage may be acceptable in certain cases where the simplicity of these elements allows for rapid solution and good contact performance. The following paragraphs related to element class and density may provide some guidance.

5.0 Coupled thermo-mechanical analysis and element density

In this model, a similar situation to the above where there is a discontinuity of temperature initially is simulated, but in this case a coupled thermal and mechanical analysis is performed. To avoid the initial value problem with shared nodes, the weld metal and the plate metal are meshed independently and then the meshes linked with constraint equations. This means that the elements start with a discontinuous temperature field that is made continuous at the start of the transient solution representing a thermal shock problem. Considering only reduced integration elements of both orders with varying mesh densities, the following models are considered.

Model 1. 4 node elements, 2*3 mesh in weld region
Model 2. 8 node elements, 2*3 mesh in weld region
Model 3. 4 node elements, 5*10 mesh in weld region
Model 4. 8 node elements, 5*10 mesh in weld region
Model 5. 4 node elements, 10*20 mesh in weld region
Model 6. 8 node elements, 10*20 mesh in weld region

The averaged stress plots are shown in figure 4 at a time of 10 seconds from the start and the comparative peak principle stress values in table 2. A good comparison is the surface stress plot that is shown in Table 3. These two tables illustrate the dangers of relying on a simple measure of stress for comparative purposes. The peak principle stress comparison indicates that the range of stresses from the coarse to the fine is fairly small and any of these analyses

may be suitable, but when comparing the distribution of a component stress we see that the lower order coarse mesh gives significantly lower stress predictions. The standard and fine higher order meshes give almost identical results indicating a 'converged' result with the standard mesh. The standard lower order mesh predicts stresses about 15% below the 'converged' values.
The outcome of this study will be a statement of the obvious to those familiar with FEA convergence checking. The higher order elements predict larger values of stress and more elements give a higher value. The assumption here is that the higher the value, the nearer to the 'real' number we will be. Other studies indicate that with some rare exceptions, this is true. The art of FEA modelling is in choosing a mesh that is economic to analyse, and give sufficiently accurate results for the purpose of the analysis. If we define an optimal mesh as one which gives a result within 2% of the ideal solution, in most practical applications the meshes chosen are sub-optimal in that the necessary mesh required for a converged solution is usually too fine to compute.

6.0 A more realistic model – pre-heating of the plate edge

As discussed above, the real welding process will pre-heat the plate edge before actual deposition of the weld. There will also be some local softening of the plate metal immediately following first contact with the weld metal before they fuse and cooling commences. This model attempts to represent this behaviour by first carrying out a coupled analysis of the plate by rapidly heating the edge of the plate up to the weld metal temperature, introducing the weld metal at this point, and allowing cooling of the complete model. The variables here are the rate of heating in the elements and the method of applying the heat. The rate of heating will depend on the detail of the real process; Total energy supplied, feed rate of the weld, energy losses through electrical efficiency and wide band radiation. Here two rates of heat have been used, sufficient to raise the local temperature to melt. Because of diffusion, the total heat introduced is different; greater for the lower flux. Again lower and higher order elements have been considered.

The outcome of this is a satisfactory model that behaves rationally. In the first stage before the introduction of the weld metal, we observe a rapid heating of the edge of the plate that diffuses into the plate. The local expansion creates stresses resisting the deformations. As the diffusion proceeds, the site of the stresses moves into the plate. The stresses will then reduce, but will not decay to zero if the heat were to diffuse completely, as a considerable amount of plastic strain will have taken place in the hot regions. Before the diffusion is complete, we remove the source of flux and introduce the weld metal at melt temperature. This is, at this stage, fully connected to the deformed plate metal. As diffusion of the heat proceeds, we see the stress field engendered by the local heating being overtaken by the thermal stresses due to differential cooling (Figures 5). The peak axial stress results are summarised in Table 4. Here we see that the stress field changes dramatically on cooling. Comparing rates of application of the pre-heat we see that for both after pre-heat and after cooling, the distribution of stress changes a little, but not the total range of stress. Comparing higher and lower order element types, we see a significant underprediction of peaks from the latter.

The difference between this and the models above is that the discontinuity stresses are minimised because we have a continuous and compatible distribution of both strains and temperature throughout the model, even when the weld metal is introduced.

7.0 Conclusions

1. The use of reduced integration elements of whatever order is recommended. Whether lower or higher order elements are used and at what density are dependant on the accuracy required and the modelling tools available. In general a larger number of simple elements are easier to use, especially if contact problems are included, but the calculation of surface stresses will always be imperfect.
2. It is necessary to model as closely as possible the real process, introducing the weld metal at appropriate conditions in the parent metal. The method is particularly unforgiving when thermal discontinuities are introduced.
3. The method developed here will be suitable for three dimensional problems where 3D diffusion takes place and weld is introduced gradually. Mesh densities indicated here will be necessary in 3D.
4. Multi-pass welds can also be simulated using this method, although a similar approach to re-heating the surface of earlier passes must be incorporated to avoid repetitive applications of thermal discontinuity.
5. A study of the effects of modelling phase change and re-crystallisation of the material on the final stress and distortion states will be necessary to complete this appraisal of necessary analysis parameters.

Table 1		
Comparison of element types on the same mesh density (element stresses in brackets)		
Model type	**Largest tensile principal**	**Largest compressive principal**
Lower order – full integration	1041 (1449)	-957 (-1294)
Lower order – reduced integration	867 (868)	-595 (-604)
Higher order – full integration	1045 (1045)	-667 (-837)
Higher order – reduced integration	875 (998)	-625 (-731)

Table 2		
Comparison of mesh densities with reduced integration elements		
Model type	**Largest tensile principal**	**Largest compressive principal**
Lower order – Coarse mesh	771.3	-601.6
Higher order – Coarse mesh	725.5	-613.6
Lower order – Standard mesh	773.4	-611.4
Higher order – Standard mesh	856.4	-620.0
Lower order – Fine mesh	775.8	-615.2
Higher order – Fine mesh	778.0	-620.2

Table 4		
Comparison of axial stresses with two rates of pre-heat and two elements types		
Model type	**Largest tensile axial**	**Largest compressive axial**
Higher order - fast pre-heat After pre-heat	81.9	-587.3
Higher order - fast pre-heat After cooling	309.7	-265.1
Higher order - slow pre-heat After pre-heat	156.9	-511.4
Higher order - slow pre-heat After cooling	337.0	-239.6
lower order - slow pre-heat After pre-heat	144.5	-386.6
lower order - slow pre-heat After cooling	260.6	-187.7

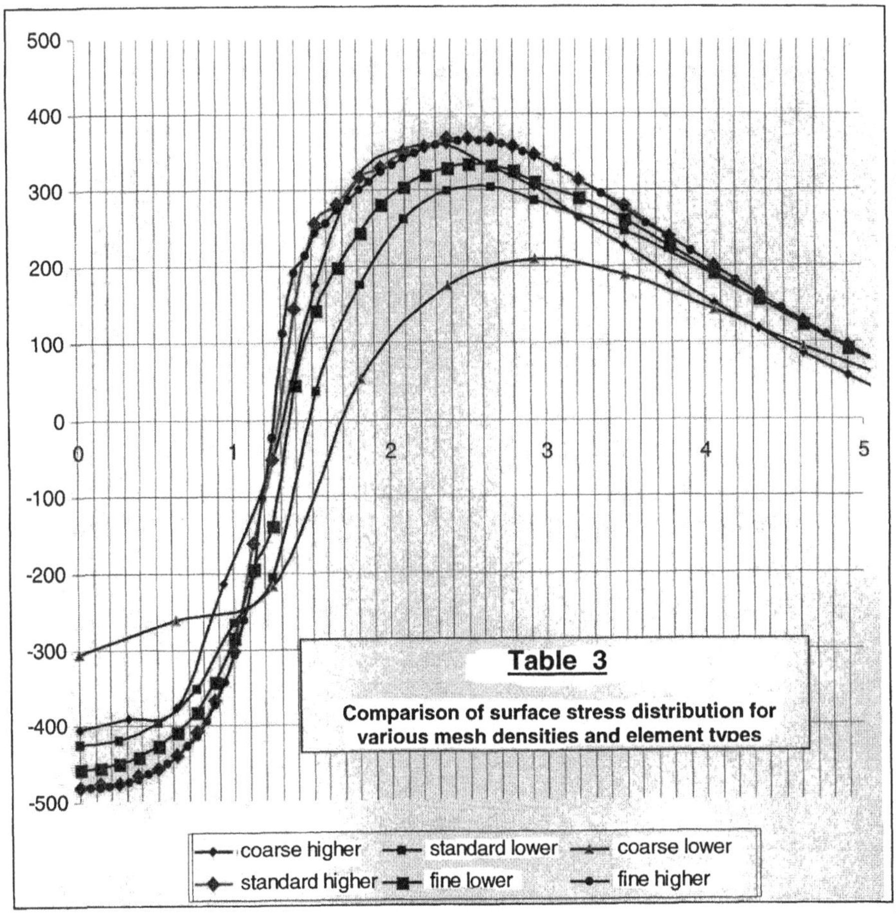
500
400
300
200
100
0
-100
-200
-300
-400
-500
0
1
2
3
4
5
Table 3
Comparison of surface stress distribution for various mesh densities and element types
coarse higher
standard lower
coarse lower
standard higher
fine lower
fine higher

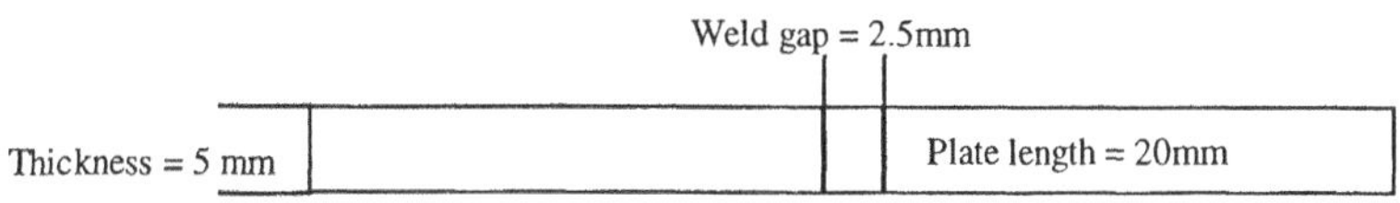

Figure 1
Geometry of problem

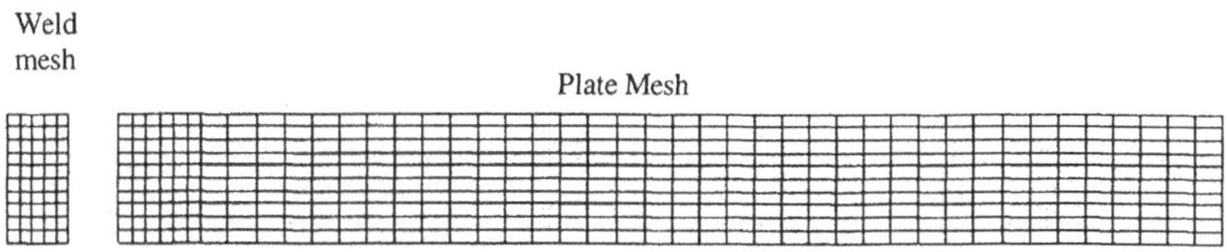

Figure 2
Mesh of quarter model
(separated for clarity)

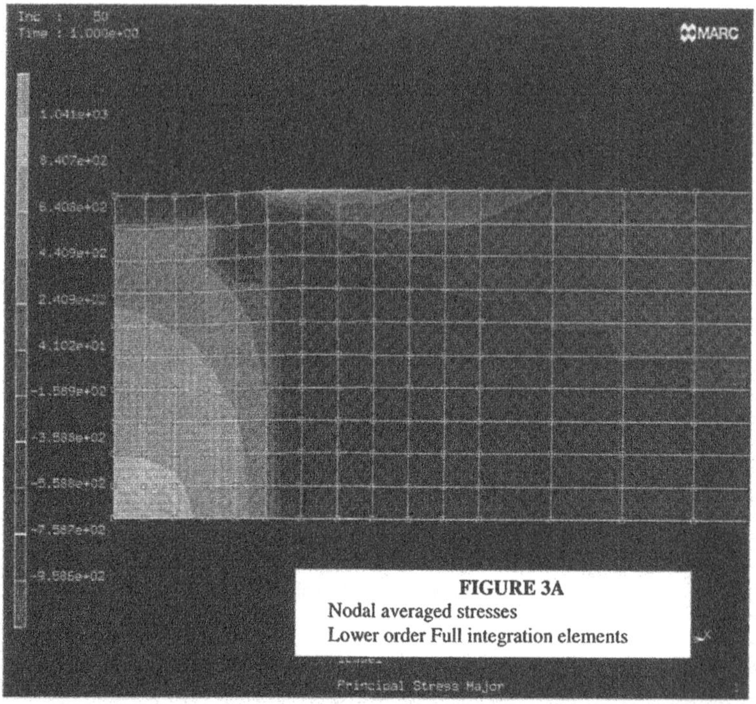

FIGURE 3A
Nodal averaged stresses
Lower order Full integration elements

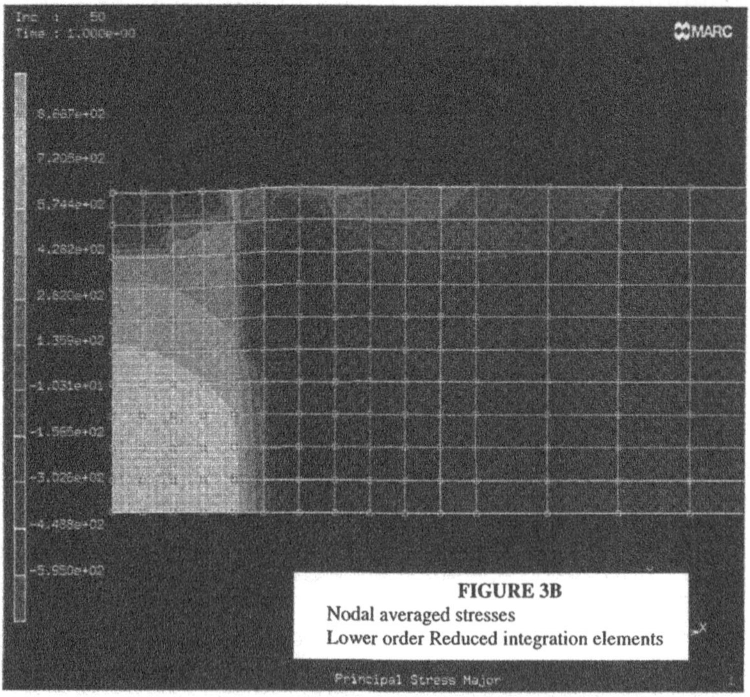

FIGURE 3B
Nodal averaged stresses
Lower order Reduced integration elements

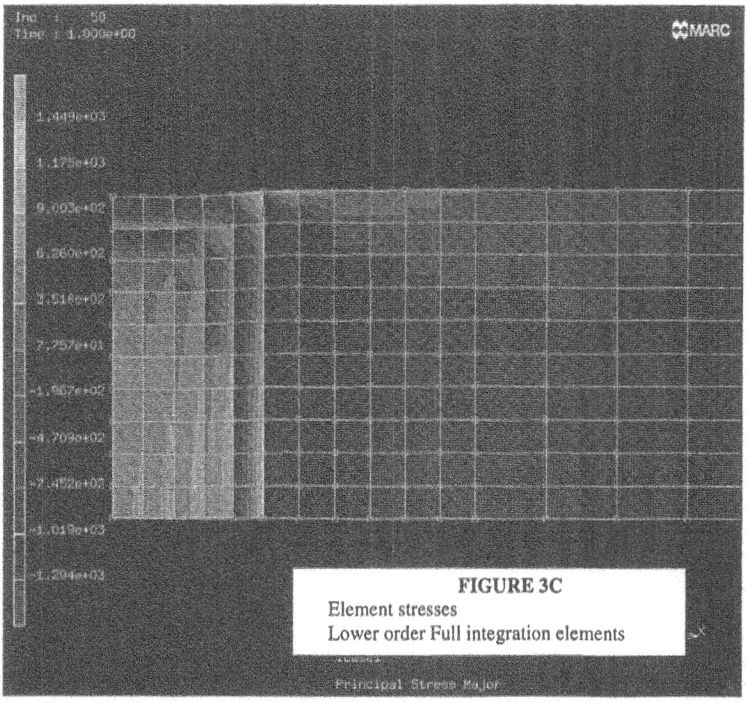

FIGURE 3C
Element stresses
Lower order Full integration elements

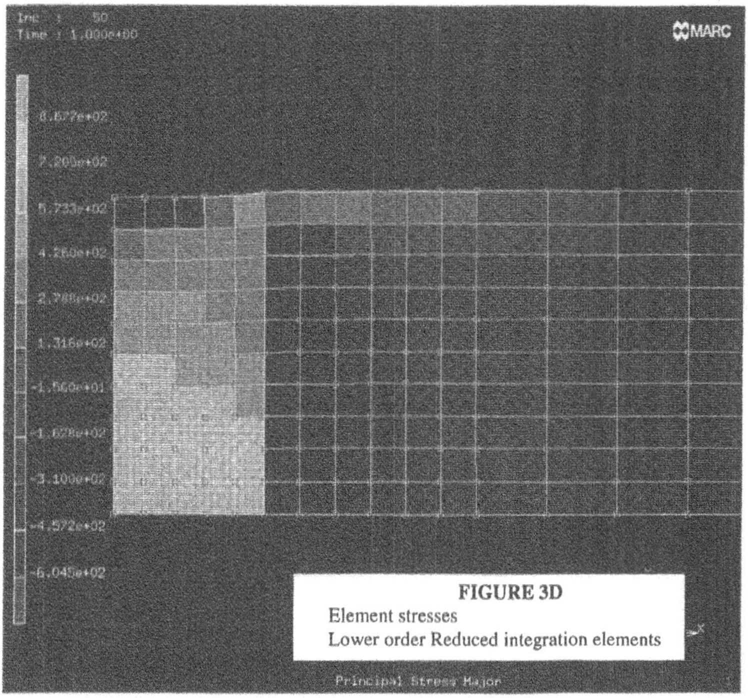

FIGURE 3D
Element stresses
Lower order Reduced integration elements

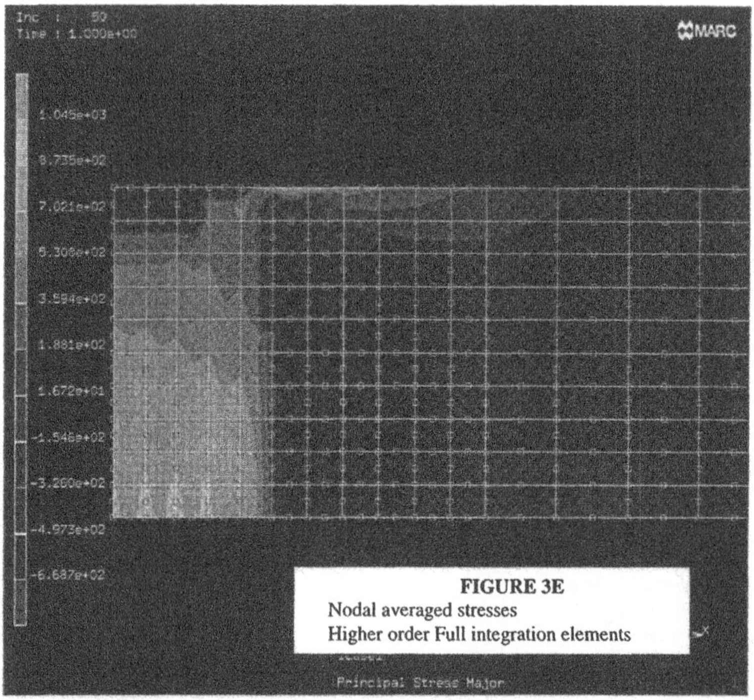

FIGURE 3E
Nodal averaged stresses
Higher order Full integration elements

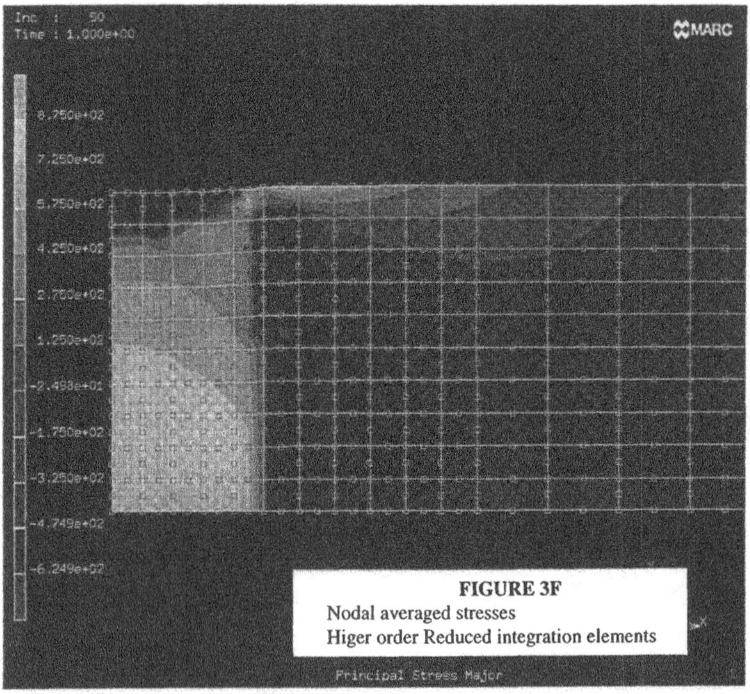

FIGURE 3F
Nodal averaged stresses
Higer order Reduced integration elements

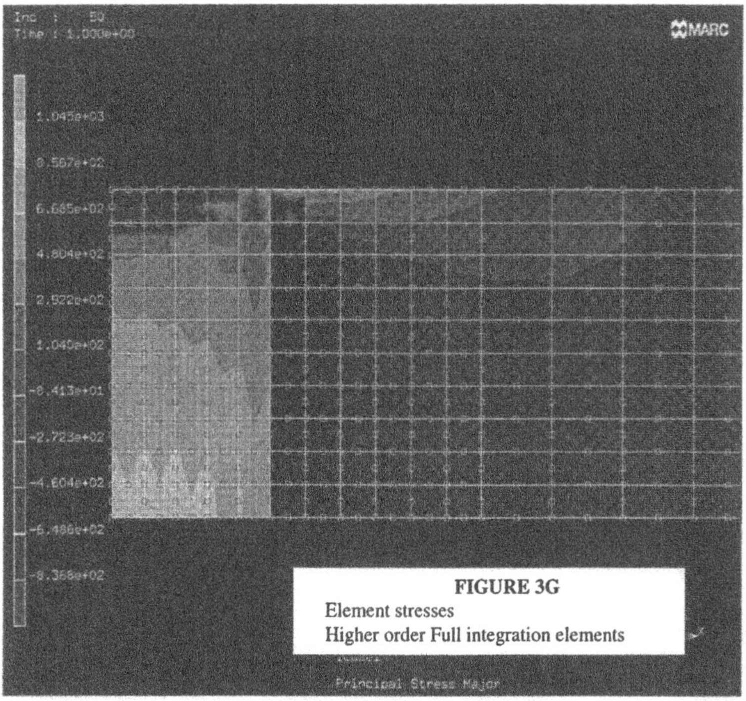

FIGURE 3G
Element stresses
Higher order Full integration elements

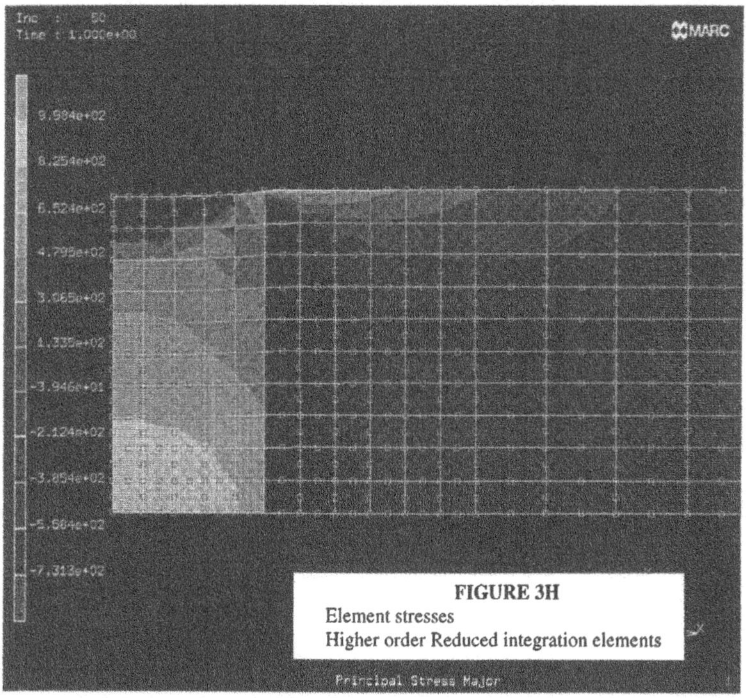

FIGURE 3H
Element stresses
Higher order Reduced integration elements

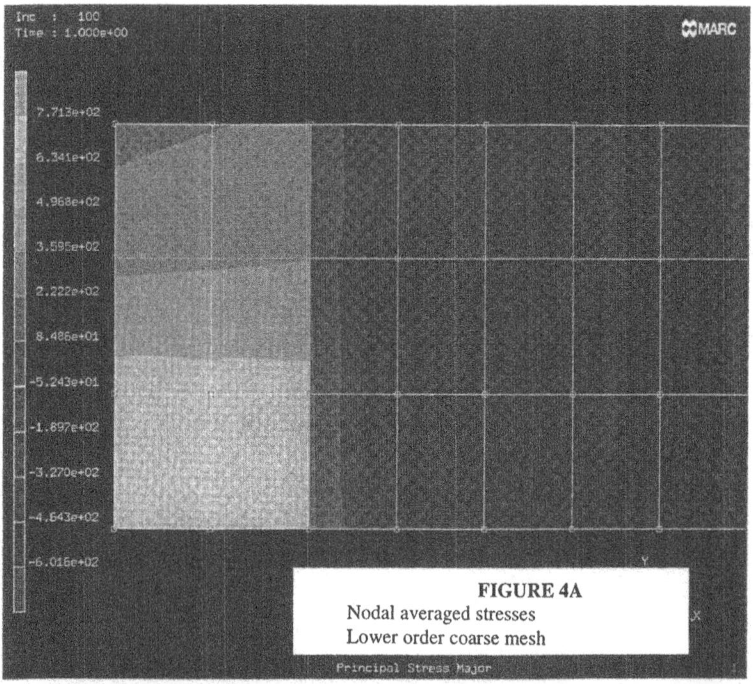

FIGURE 4A
Nodal averaged stresses
Lower order coarse mesh

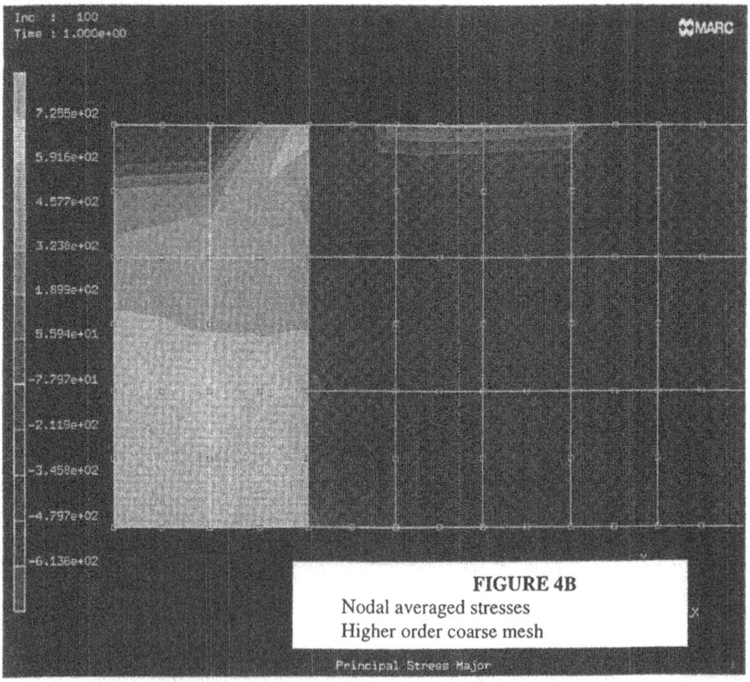

FIGURE 4B
Nodal averaged stresses
Higher order coarse mesh

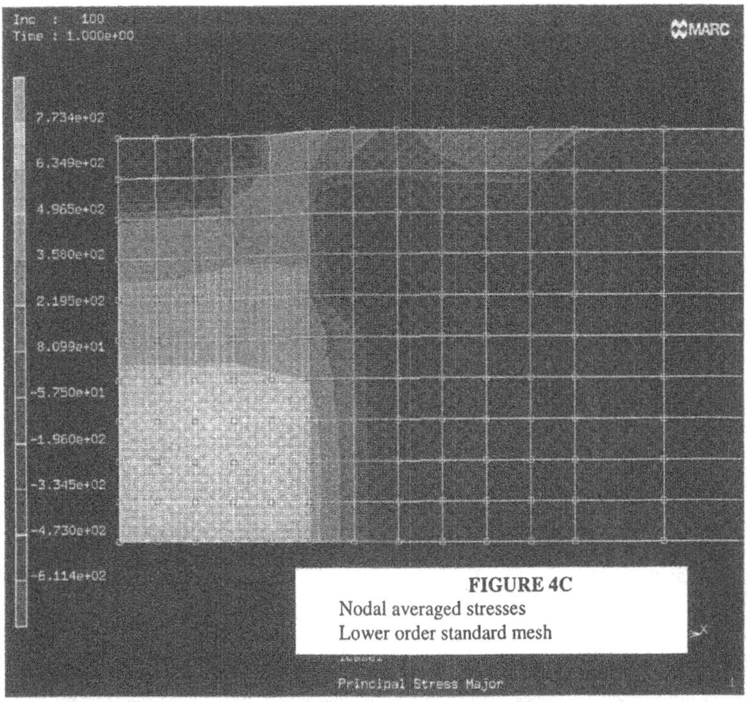

FIGURE 4C
Nodal averaged stresses
Lower order standard mesh

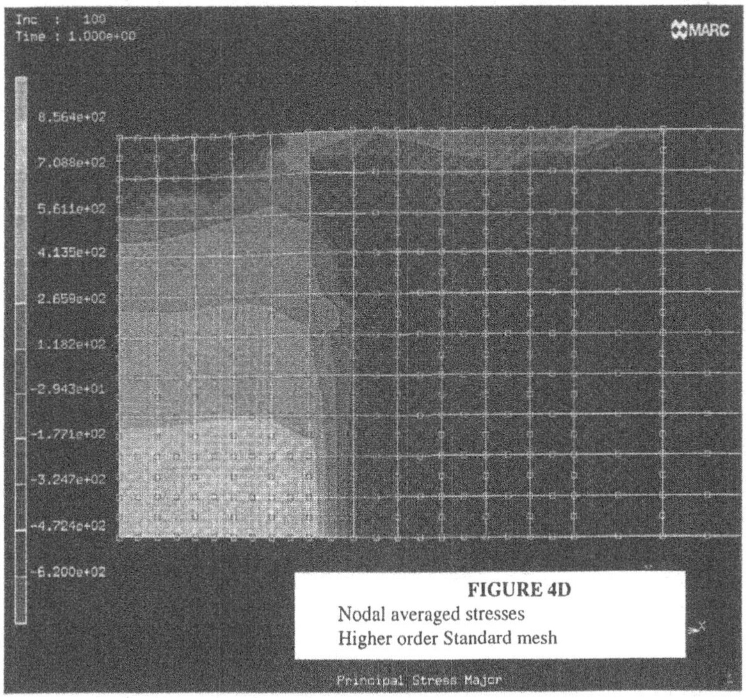

FIGURE 4D
Nodal averaged stresses
Higher order Standard mesh

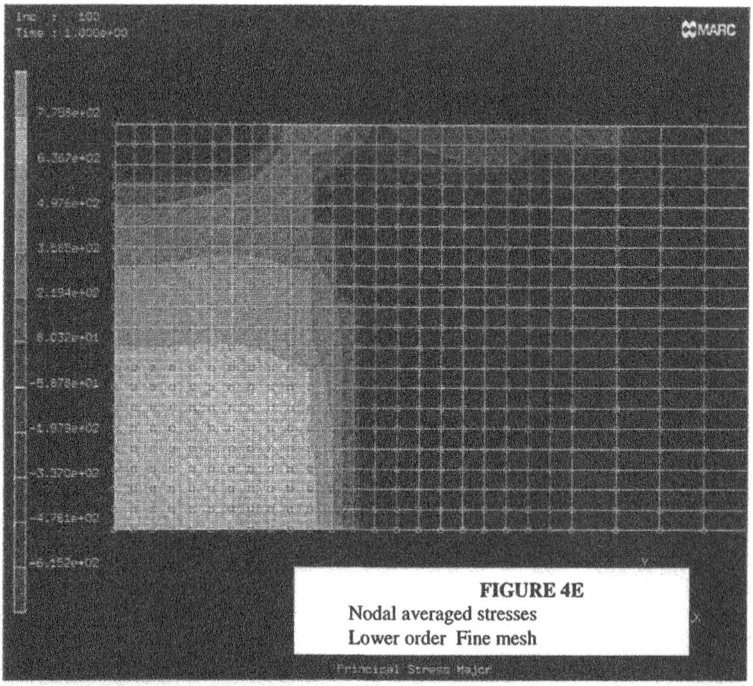

FIGURE 4E
Nodal averaged stresses
Lower order Fine mesh

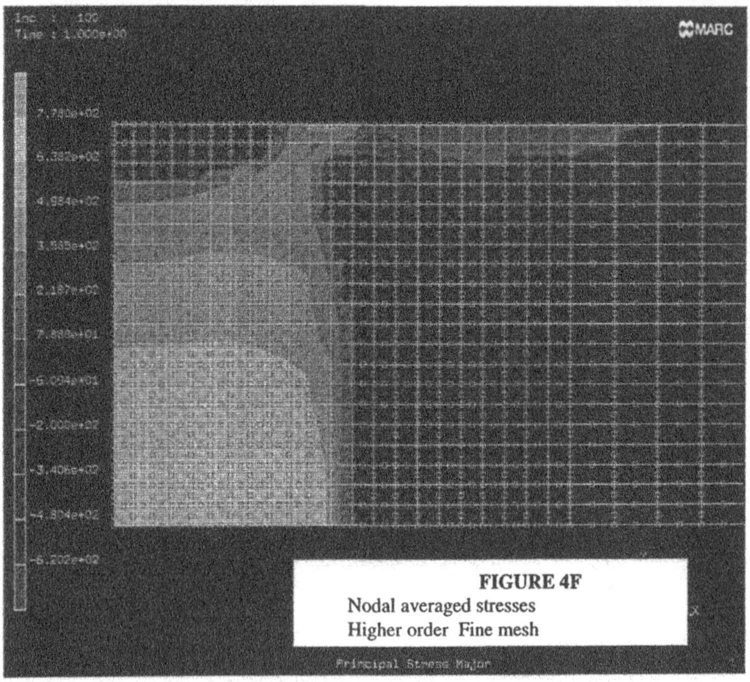

FIGURE 4F
Nodal averaged stresses
Higher order Fine mesh

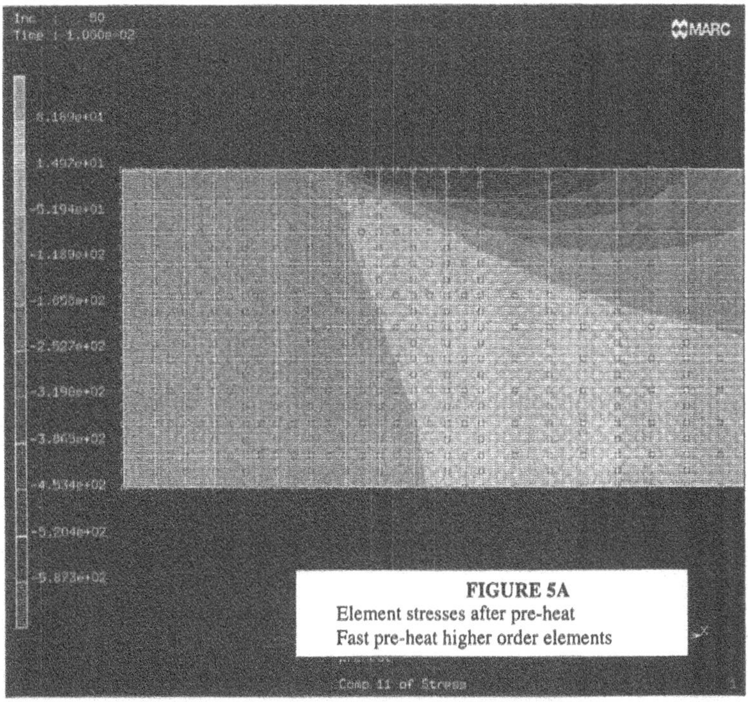

FIGURE 5A
Element stresses after pre-heat
Fast pre-heat higher order elements

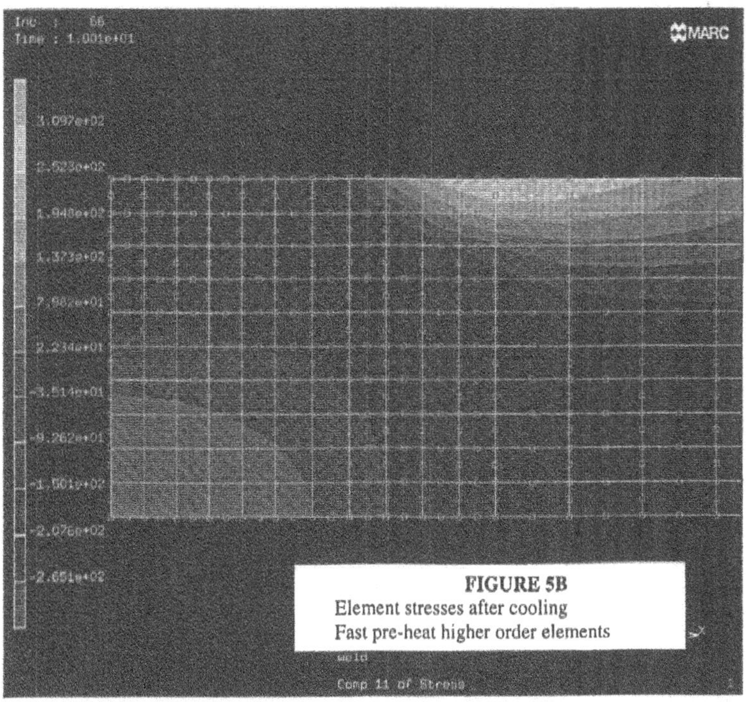

FIGURE 5B
Element stresses after cooling
Fast pre-heat higher order elements

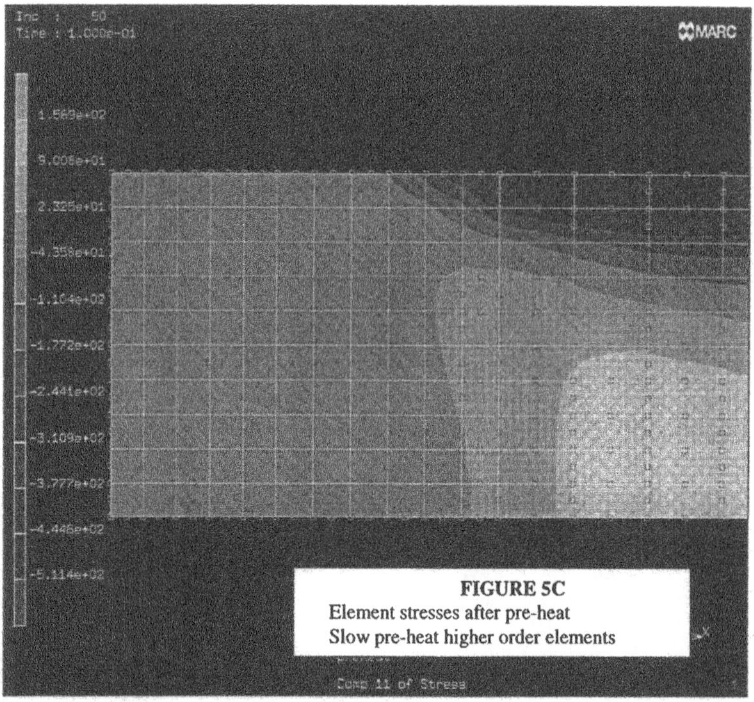

FIGURE 5C
Element stresses after pre-heat
Slow pre-heat higher order elements

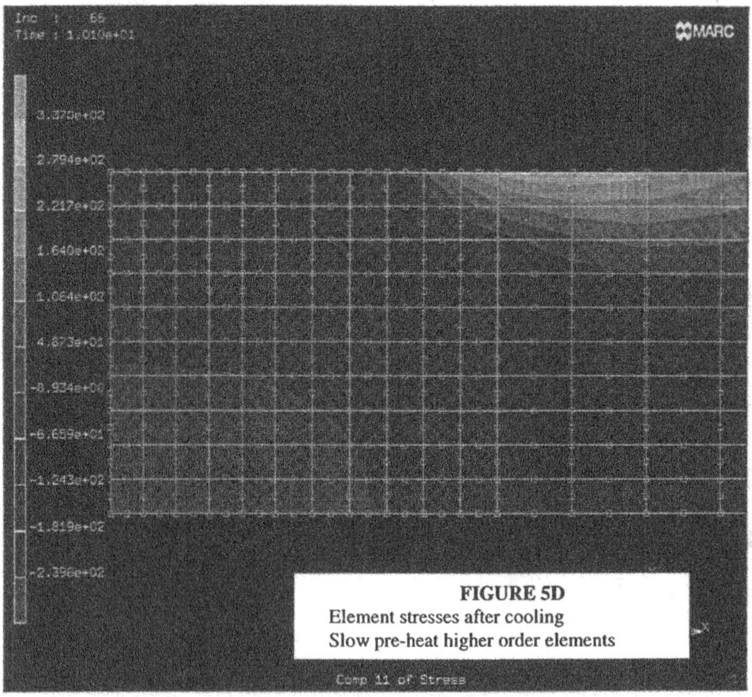

FIGURE 5D
Element stresses after cooling
Slow pre-heat higher order elements

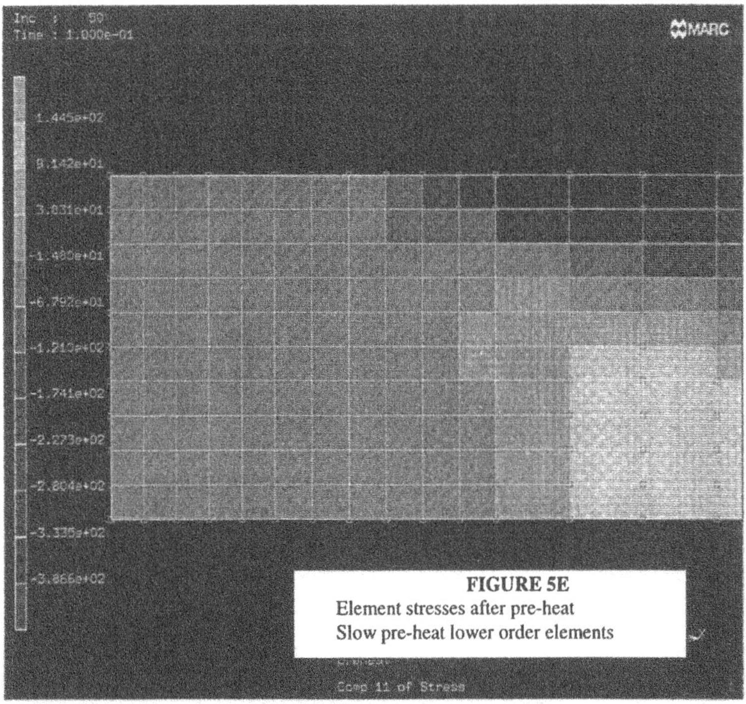

FIGURE 5E
Element stresses after pre-heat
Slow pre-heat lower order elements

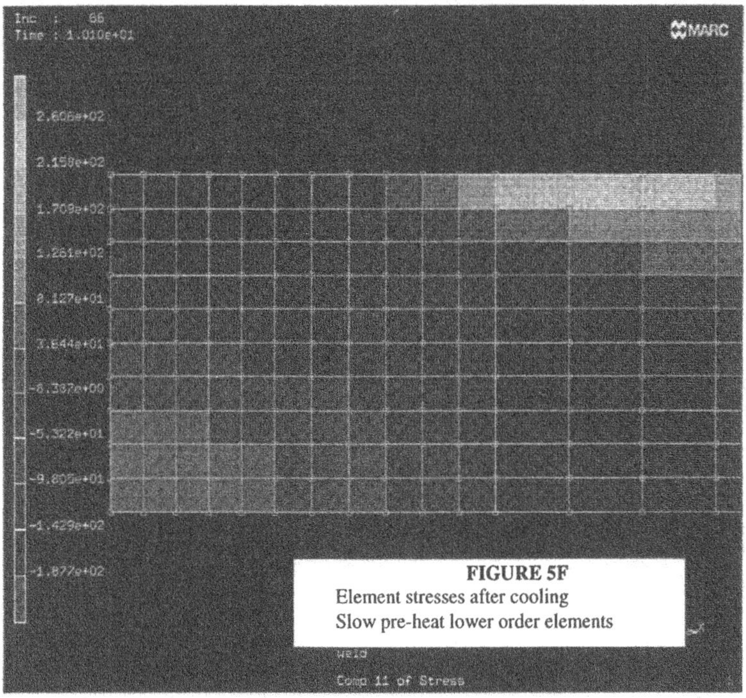

FIGURE 5F
Element stresses after cooling
Slow pre-heat lower order elements

Authors' Index

B

Bouchard, P J 55–78

C

Carter, T 45–54

D

Dye, D 33–44

E

Emmerhoff, O J 79–88

G

George, D 55–78

H

Hunziker, O 33–44

J

Josefson, B L 1–20

K

Kohlhöfer, W 89–98

L

Leggatt, R 21–32
Lindgren, L-E 1–20

M

Mangialenti, G 45–54
Midling, O T 79–88

O

Olden, E 21–32
Oosterkamp, L D 79–88

P

Penny, R K 89–98

R

Ramsay, A C 99–118
Reed, R C 33–44
Roberts, S M 33–44

S

Smith, D J 55–78
Stone, H J 33–44

W

Watson, C 55–78

Printed and bound by CPI Group (UK) Ltd, Croydon, CR0 4YY

13/09/2026

14961031-0006